Thomas Hames Pearmain, Cresacre George Moor

The Analysis of Food and Drugs

Part II. - Chemical and Biological Analysis of Water

Thomas Hames Pearmain, Cresacre George Moor

The Analysis of Food and Drugs
Part II. - Chemical and Biological Analysis of Water

ISBN/EAN: 9783744643610

Printed in Europe, USA, Canada, Australia, Japan

Cover: Foto ©berggeist007 / pixelio.de

More available books at **www.hansebooks.com**

THE ANALYSIS OF FOOD
AND DRUGS.

Part II.
CHEMICAL AND BIOLOGICAL ANALYSIS
OF WATER.

BY

T. H. PEARMAIN and C. G. MOOR, M.A. (Cantab.),

MEMBERS OF THE SOCIETY OF PUBLIC ANALYSTS,

AUTHORS OF

'A MANUAL OF APPLIED BACTERIOLOGY,' 'AIDS TO THE ANALYSIS OF FOOD AND DRUGS,

ETC.

LONDON:

BAILLIÈRE, TINDALL AND COX,

20 & 21, King William Street, Strand.

[PARIS AND MADRID.]

1899.

PREFACE.

WATER is not technically included among foods or drugs, but it enters so constantly into the ordinary practice of an analyst whose work lies in their examination that it appears to us desirable to include it in the present manual.

It has been impossible in the space at our disposal even to refer to many of the enormous number of processes which have been proposed for the estimation of the various substances contained in waters. In making a selection, we have chosen those which by actual prolonged use we have found to be convenient, and, within practicable limits, accurate.

The information in regard to the suspended matter in water is illustrated by original drawings, as the recognition of the larger organisms is easier than that of the bacteria, and might be practised with advantage more often than is usual. The majority of the drawings have been made for this work from her own preparations by Miss Ethel Smithwhite.

We are indebted to Mr. Wolf Defries for criticism, more particularly in regard to the circumstances which affect the significance of results and to the 'duty' of several methods of examination. Although the views which Mr. Defries holds in regard to these matters, both from the chemical and the bacteriological standpoint, are stronger than those which it is usual for analysts to express, we have adopted them with the greater satisfaction because we do not think that the limitations peculiar to the examination of water are as clearly recognised as they should be. We have also to thank Dr. E. H. Cartwright for his kindness in reading the whole of the proofs, and to many other analysts for much information as to their personal methods.

T. H. P.
C. G. M.

4, DANE'S INN, STRAND, W.C.

CONTENTS.

THE CHEMICAL AND BIOLOGICAL ANALYSIS OF WATER.

I. INTRODUCTION.

It is now universally acknowledged by sanitarians that a proper supply of unpolluted drinking-water is one of the most essential conditions to ensure the health of the community. This being so, it is one of the first duties of the Sanitary Authorities to see that this is provided to every house under their jurisdiction. Most of our large towns are provided with a more or less satisfactory public water-supply, but the larger proportion of the population, who are resident in villages and isolated dwellings all over the country, are dependent upon water-supplies derived generally from shallow wells, which in many cases are grossly polluted. This state of affairs is all the more serious, in view of our certain knowledge of the conveyance of disease by means of polluted water-supplies. One is frequently met with the retort from ignorant persons that such waters have been drunk without harm from time immemorial. Although water contaminated with sewage has been, and is still, drunk by a large number of people apparently with impunity, yet such dangerous waters may at any moment become positively injurious to the health of the persons using them. It is now well known that polluted water is the most important agent in the convey-

1

ance of typhoid fever. The slightest contamination of a water-supply with the dejecta from a single case of enteric fever has, in many well-authenticated instances, caused widespread epidemics of the disease amongst those persons who have used the contaminated water-supply, as witness the recent epidemics at Worthing and Maidstone.

The object of water analysis is, as a rule, to ascertain the suitability of a water from a particular source for drinking and other purposes. In judging of the suitability of a particular supply, it is advisable to ascertain its history as far as possible—that is, the character of the geological strata from which it is derived, the relative positions of possible sources of pollution, such as cesspools, churchyards, stables, etc. This information is frequently withheld by persons submitting a sample for analysis for fear of biassing the analyst's opinion, and thus information that might have been of use is often wanting.

In the absence of the history of a water, we have generally to rely on the information we are able to gain from the three methods of examination, which we place in the order of the importance we attribute to them :

(1) Chemical analysis.
(2) Biological examination.
(3) Physical examination.

The first point to be clearly grasped is that neither any one, nor the whole, of these methods, applied to a single sample or to a few sporadic samples of water, can warrant the fitness of a water-supply for drinking purposes. The analysis of such sample or samples may disclose sufficient quantities of substances or bodies of suspicious origin to show that the water is unfit for consumption, or, at least, that it is either derived from sources liable to convey infective material or contains poisonous substances. It may, however, give no such positive results, and this

circumstance may be due either to the fact that the water
is either pure in itself and not open to pollution, or to the
fact that at the time at which the sample was taken no
pollution happened to be drawn or forced into the supply.
The latter interpretation is very frequently the correct one;
and the danger which attends it, if it be not recognised by
the Sanitary Authorities, is that due to false confidence,
and has been responsible for some of the most serious
water-borne epidemics. The means of protection which
are available through analysis are due to two facts. When
infective organisms are introduced to a water initially
proper for consumption, these organisms do not come
alone, but are usually accompanied by chemical substances
and by other organisms, which alter the chemical and
biological composition of the water in a manner to be
detected by analysis. When no substances which are
infective or liable to convey infection are introduced into
a water-supply, its composition is usually found to be
practically constant. It is possible, by taking observations
over a sufficient period, to determine whether the composition of a water-supply is constant or variable; and
before it can be warranted as safe, it must by such
systematic and continued examination have been shown to
be of practically constant composition in all its factors.

The first, and beyond comparison the most important,
fact in the analysis of any water-supply for sanitary purposes is that its examination at short periodical intervals is,
without reservation, the only means at disposal for obtaining the protection which technical examination can give.
It may be said, indeed, that even a partial examination conducted under rigidly constant conditions every day, or even
every week, is preferable to a complete examination made
once a quarter, and would yield a far greater increase of
security to those who use the water.

Each of these methods of examination of water has its own special importance, but it may be as well to indicate the information which is gained by these methods of examination. Undoubtedly the chemical examination is usually the most important.

It permits a far larger number of independent factors to be accurately estimated; and although the consumption of the substances discovered may in itself be indifferent to health, the chance of detecting variation in composition, which is the most important sign of danger in waters which may sometimes or usually be safe, is considerably greater than by other methods. Its least merit is the certainty by which directly poisonous, as distinguished from infectious, material is detected; for such material is of rare occurrence. In the same way, but not for the same reason, it may be said that the least merit of bacteriological examination is the direct detection of specific infective organisms. Where such organisms are detected the supply is, of course, to be condemned. But it must be remembered that infective organisms are not dissolved in the water and uniformly distributed over it. They are solid bodies floating in it, and as only a very small sample of the total supply is examined, they may be present in the supply and absent from the particular samples chosen. They are also in all cases accompanied by other organisms, some much better capable of growing in water; and when they reach the analyst, they may be so far attenuated, that they fail to develop in his media, and though present, may escape recognition. The chief merits of bacteriological examination are two. On the one hand, it furnishes a factor capable of quantitative estimation, by which variations in the composition of the supply may be detected. On the other hand, it enables organisms which, though themselves indifferent, are usually introduced with infective material to be detected. It also enables us to control the working of water-filters.

Respecting the information that we may gain from the microscopic examination of suspended matters, it must be remembered that while we may sometimes find such bodies as epithelial scales, undigested muscular fibre, and the like, which may lead us to suspect pollution with sewage or other putrescible matter, yet in most cases of waters from towns such substances will be filtered out of the water ; this will also happen in the case of well-waters, so it is only occasionally that this method of examination will render assistance to the analyst.

The microscope also permits the recognition of organisms different from, and of larger size than, those which are known to be infective. Some of these are directly diagnostic of the sewage which is the ordinary vehicle of infective material, and where the supply has been filtered, the presence of any such larger organisms is evidence that the filter would have been unable to arrest the more minute organisms of infective disease, and that the protection intended to be obtained by the filtration has at least, for the time being, been illusory.

There is no great difficulty in performing the operations involved in water analysis, but the power of correctly interpreting the analytical results is only to be attained by very considerable amount of training and practice.

A satisfactory drinking-water should be practically free from organic matter, and should neither contain, nor have the power of dissolving, poisonous metals, and should be free from excessive hardness. Regarding the first of these requirements, freedom from organic impurity, it was clearly recognised as dangerous to drink water polluted with excrementitious matter long before the precise nature of the danger was known.

II. CLASSIFICATION OF WATERS.

THE term pure cannot be applied to natural waters in its strict sense, but only relatively, since absolutely pure water does not exist in nature. All potable waters are contaminated with impurities to a greater or less degree ; and it is through the nature of these impurities that the analyst is assisted to form an opinion as to the previous history of the water and its probable suitability or unsuitability for potable purposes. Owing to the remarkable solvent power possessed by water, the number and variety of substances which it can take into solution greatly exceed that of any other liquid. In addition to its great solvent power for many saline and gaseous bodies, it has considerable solvent power for many of the constituents of the rocks and strata composing the earth's crust. We may classify the principal constituents and impurities of natural waters as follows :

GASEOUS.—Carbonic acid, oxygen, nitrogen.

SOLIDS (*Mineral*). — The carbonates of calcium and magnesium, sulphates of calcium, magnesium and sodium, chlorides of sodium and potassium, silicates.

SOLIDS (*Organic*).—The products of the decomposition of animal and vegetable matters.

In addition to these, waters are liable to contain more or less insoluble and suspended matters, consisting of mineral matters, such as sand, clay, oxide of iron, chalk, etc., or of living and dead vegetable and animal matters consisting of algæ, moulds, bacteria, hairs, cellular tissue, fibres, insects, infusoria, epithelial scales, fæcal débris, etc.

Natural waters can be divided into five classes :

Rain-water.—This includes all water derived direct from the atmosphere, such as rain, snow, hail, and dew.

Surface-water.—Waters derived from lakes, ponds, rivers, and streams.

Subsoil-water.—Waters percolating through the soil or rocks, at no great distance from the surface, and lying between it and the impermeable stratum.

Deep Well or Artesian Well Water.—Waters obtained by means of borings at a considerable distance from the surface, and coming from below the impermeable stratum.

Sea-Water.

Generally speaking, we may arrange the various sources of supply of potable water in the following order with regard to purity :

Rain-water (collected under proper conditions).

Deep spring and artesian well water.

Upland surface and moorland waters.

Subsoil-water (if collected distant from any source of pollution).

River-water.

Land springs.

Surface-water from cultivated lands.

RAIN-WATER.

This is the purest form of natural water, if collected under proper conditions in the open country. Rain-water contains only a trace of solid matter, and is therefore perfectly soft; but it contains frequently large quantities of ammonia, together with traces of nitrates and nitrites. The traces of solid matter are, of course, derived from the dust and other matters contained in the atmosphere. The rain-water from the neighbourhood of large towns is very impure, containing as it does large quantities of soot, sulphurous acid, and other products of combustion.

Rain-water, owing to its soft character, has a great tendency to act upon leaden pipes and cisterns, becoming charged with this poisonous metal. In districts near the sea rain-water will be found to contain very appreciable amounts of chlorides. The River Pollution Commissioners, in their Report, give a lengthy table of analyses of seventy-eight samples of rain-water, and of eight samples as ordinarily collected and stored in tanks. The following are the averages of their results :

				In grains per gallon. Fresh rain-water.		Tank-water.
Total solids ...	...	...	...	2·76	...	16·8
Chlorine	...	...	...	0·004	...	0·78
Hardness	...	...	...	0·42	...	7·9
Nitrogen as nitrates	...	...	0·43	...	0·78
Free ammonia	...	...	...	0·035	...	0·08

Levy has found that in dew, fog, and snow-water there is more ammonia and less nitric acid than in rain-water.

Muntz has shown that the rain-water of the tropics is richer in ammonia than that of temperate climates, and that town air contains more ammonia than does country air.

Where rain-water has to be collected from the roof for drinking, it may be necessary to calculate how much can be expected from this source. To take an example, if the rainfall were 24 inches per annum, and the roof (which should be slate or tiles) were 60 feet by 20 feet, we should in a year get 60 feet by 20 feet by 2 feet, or 4,800 cubic feet, or 28,240 gallons, theoretically. In practice, however, the loss due to evaporation is considerable; and to obtain the water fit to drink it would be essential to employ a rain-water separator, an ingenious arrangement which discards the first few gallons of roof washings, and, after the roof has been well cleansed, the rest of the shower is diverted into the store tank for use.

In the above case, if we imagine the cottage to contain six persons, and that one-third of the water is lost, we shall have 18,826 gallons per year, or about 8·6 gallons per head per day—not a very liberal allowance.

Where rain-water is to be collected for use, the keeping of pigeons is, for obvious reasons, a very undesirable practice.

SURFACE-WATER.

As already mentioned, the whole of our potable water is derived originally from the atmosphere in the form of rain. Part of this sinks into the earth, and the remainder runs over the surface of the land in the form of streamlets, which unite to form rivers, ponds, lakes, etc. The water as it flows over the surface takes up more or less suspended and dissolved matters, the former being found in larger amounts when the rainfall is plentiful. The wearing action of water is largely dependent on the nature of the suspended matter that it conveys down from the upper levels of the watershed. When the tributaries conveying the waters from districts of varying characters become mixed, the nature of the suspended and dissolved matters may become largely modified. When the streams from the higher levels of a watershed, more or less in the form of rapid mountain torrents, gather together to form rivers, the rapidity of the current naturally becomes slower as they reach the lower levels, and as a result much of the suspended matter is deposited. The water of streams and lakes is in part derived from springs, which in many cases come from great depths, and thus may introduce into solution substances which are not present in surface-water proper.

By the joint action of air and water the hardest rocks undergo a weathering or disintegrating process. By the exposure of the surface of the rocks to the alternate freezing and thawing of the water contained in the minute interstices, the superficial layers become broken down, thus enabling the water to exercise a solvent action upon the soluble constituents, the insoluble portion being conveyed down to the lower levels in the form of gravel and mud.

The Igneous, Metamorphic, Silurian, Cambrian, and Devonian rocks, which are so largely exposed in the western parts of England and Wales, yield a very considerable amount of surface-water. The water so yielded is generally exceedingly soft and of high organic purity.

Rivers are very liable to pollution, as they form the natural destination for the drainage from the land situated on each bank. From the earliest times inland towns have always, when possible, been established upon the banks of a river or stream, in order to be within easy reach of an abundant supply of drinking-water, and for the convenience of commerce. As the populations increased, the amount of water used and fouled became greater and greater, owing to the soil becoming saturated with excremental matter due to the earth method of disposal which was in vogue for so many years.

SUBSOIL-WATER, OR GROUND-WATER.

Subsoil - water is that which has percolated the strata immediately below the soil, and is generally easily obtained by means of shallow wells. Subsoil-water is one of the most frequent sources of supply, particularly in villages, but it is exceedingly liable to pollution, especially in the vicinity of houses, farms, etc. The amount of water which enters the soil depends upon the permeability, and varies with the nature of the strata, gravel soils being the most and clayey soils the least permeable. The quality of subsoil-water also varies with the nature of the subsoil and the proximity to houses. Subsoil-water from the chalk, lias, oolite, and limestone districts will be hard, but that supplied from the older formations, such as the Yoredale and millstone grits, as also that from sands and gravels generally, will be soft if uncontaminated. Water collected by means of shallow wells in the proximity to human habitations is often most dangerous to health, owing to its receiving the leakage and soakage from privies, drains, and cesspools, although, notwithstanding the presence of these dangerous polluting matters, it is normally bright and palatable.

DEEP OR ARTESIAN WELL WATER.

Deep borings, or artesian wells, particularly those which pass through impervious into pervious water-bearing strata, are very important sources of water from the hygienic point of view, since many of these wells provide practically unlimited amounts of water of almost absolute organic purity. The geological conditions do not always, however, favour this type of well. The older rocks, such as the Devonian, millstone grit, magnesian limestone, and coal measures provide but little water by this means. The new red sandstone, oolites, chalk, lias, and greensands are the great water - bearing strata. These formations contain vast volumes of water stored up in their pores and fissures, and are capable of providing practically unlimited supplies of drinking-water of the highest purity, and it is a matter of surprise that greater use is not made of this source of supply. Fissures may, however, establish connection between the supply and the surface water, and in that case dangerous pollution may arise. The best precaution, and often the only one which is possible, against this hidden danger is the systematic examination of the water, for its composition will be normally stable, and substantial variations will indicate the intrusion of a foreign source. The Rivers Pollution Commissioners found the deep-well waters from the chalk 'almost invariably colourless, palatable, and brilliantly clear. The chalk formation constitutes magnificent underground reservoirs, in which vast volumes of water are not only rendered and kept pure, but stored and preserved at a uniform temperature of about 10° C., so as to be cool and refreshing in summer, and far removed from the freezing-point in winter.'

SEA-WATER.

Sea-water is but of little importance from the hygienic point, but is of great interest chemically. Sea-water taken from the surface of the ocean ranges in specific gravity from 1022·0 to 1029·0, pure water being 1000·0, and contains from 3·3 to 3·9 per cent. of dissolved solids. The water of narrow seas may contain less than this: thus the Baltic only contains 1·8 per cent. of solids. The water of inland seas or lakes may, however, be heavily loaded with dissolved solids; for instance, the Dead Sea contains 24·0 per cent. of solids.

The solids of sea-water consist in the main of common salt (sodium chloride). According to Dittmar, the solids of sea-water on the average have the following composition:

Chloride of sodium	...	...	77·7	per cent.
Chloride of magnesium	...	...	10·9	,,
Sulphate of magnesium	...	...	4·7	,,
Sulphate of calcium	...	...	3·6	,,
Sulphate of potassium	...	...	2·5	,,
Bromide of magnesium	...	...	0·2	,,
Carbonate of calcium	...	...	0·8	,,

III. COLLECTION OF SAMPLES.

THE greatest care should be taken in collecting samples of water for the purpose of analysis, in order to secure average and representative samples of the supply under examination. The amount of water required for an ordinary sanitary analysis of water is half a gallon, and the most convenient receptacle for containing the samples are the stoppered 80 oz. 'Winchester Quart' bottles, which are readily to be obtained at any chemist's shop. Corked bottles should never under any circumstances be used. If a full analysis is required, as, for instance, in the case of judging the suitability of water for brewing or boiler purposes, a larger quantity of water is required, say, from one to two gallons. The bottles used for collecting the samples must first be thoroughly cleansed by washing with hot soda to remove traces of grease, and then with a little dilute hydrochloric or sulphuric acid, and must then be well rinsed with water until the last traces of acid are completely removed. It is preferable, whenever possible, to use new bottles, but whether the bottle is new or not, it must be well rinsed with water several times, and lastly with distilled water and well drained. It is safest never to use bottles which have contained organic fluids.

If the sample of water to be examined is to be taken from a pump or hydrant, it should be allowed to run to waste for a short time before collecting. In the case of tap-water, it should also be allowed to run for a time before taking a sample, unless it is desired to examine the water for the presence of lead, in which case it may be advisable to take the first runnings, so as to see whether it is able to dissolve the metal through standing in the pipes. In taking samples from a river, streams, or lakes, care should be taken to avoid any floating scum, or the stirring up of mud from the bottom, and the sample should be always taken away from the banks. In taking samples from public supplies, they should be collected from a hydrant in direct connection with the main, and not from cisterns or other storage receptacles.

No sealing-wax, linseed paste, or other form of sealing material should be put on the stoppers of water samples, owing to the danger of contaminating the samples, but the stopper should always be covered with a clean piece of parchment paper; the string used in fastening this can be sealed with sealing-wax if desired. The bottles after labelling can then be packed in straw or other suitable packing material, and enclosed in hampers or boxes for carriage. Water samples should be sent off to the analyst with the least possible delay, so that the analysis can be put in hand as soon as possible. This is very important, as waters, particularly polluted ones, are liable to undergo chemical changes on keeping. Where it can be done, there is advantage in packing the sample in ice, and so keeping it until examination. This may not absolutely prevent change, but in so far as change is due to bacterial action, it at the least very materially retards it. It is also of the greatest importance that a proper record should be sent of the date of collection, preferably upon a label upon the bottle itself, of the source of the water, whether derived from pump, river, etc., particulars in reference to sources of possible pollution, such as proximity to cesspools, stables, sewers, or manufacturing establishments. In the case of wells the depth should be given, and also the character of the local geological conditions should be given whenever possible. Since the object of a water analysis is most frequently to determine as to whether it is fit for domestic purposes, too much importance cannot be insisted upon as to the necessity of supplying the analyst with these particulars as to its history. It is often difficult to give a correct opinion as to the purity or safety of a water without the knowledge of these facts.

It is sometimes thought prudent by persons or authorities seeking the advice of an analyst to suppress these particulars, on the ground that they will bias his judgment. It is for this reason, among others, that we attach the utmost importance to using all methods whereby those concerned may realize how very indirect is the bearing of the information which analysis yields upon the sanitary quality of a water. No analytical figure (poisonous metals apart) expresses any circumstance which in itself has the faintest bearing on health. The figures of a water put together provide, nevertheless, the most cogent means of forming a judgment on its character, provided the source of the water be known ; but they may be misleading in the absence of such knowledge, because the inference which a given set of figures may justify in regard to a water of one origin may be entirely inapplicable to a water of another origin.

When a bacteriological examination is needed upon a sample of water, special conditions and precautions have to be adopted for its collection ; these will, however, be dealt with under the section dealing with the biological examination of water. For the ordinary chemical examination of a sample of water for sanitary purposes, it is necessary to have the following data, for expressing an opinion as to its freedom or otherwise from pollution with sewage, or other dangerous putrescible matter, and general suitability for domestic purposes :

Physical characters and nature of sediment, if any.

Total solid matter.   Loss of same on ignition.

Chlorine.

Hardness.

Phosphates and poisonous metals.

Saline ammonia.

Albuminoid ammonia.

Amount of oxygen consumed for oxidation of contained organic matter.

Nitrogen present as nitrates and nitrites.

Statement of the Results of Analysis.

The results of analysis are still commonly expressed in grains per gallon, *i.e.*, in parts per 70,000, since there are this number of grains in the gallon.   Some analysts prefer to report the solid constituents as grains per gallon, and those representing the organic impurity in terms of parts per million (=milligrammes per litre).

The method of reporting in parts per 100,000 (= milligrammes per 100 c.c.) has the advantage that the results, are at once comparable with analyses made abroad.   This method of stating analytical data in connection with water analysis is in common use on the Continent, and is rapidly coming into use in this country.   Results expressed as grains per gallon can be converted into parts per 100,000 by dividing by 7 and multiplying by 10, whilst multiplying by 7 and dividing by 10 converts parts per 100,000 into grains per gallon (=parts per 70,000).

When the results are desired in terms of grains per gallon, it is convenient to measure 70 c.c. for each estimation or a multiple of this quantity by means of a special pipette. This quantity (70 c.c.) represents a 'miniature gallon,' and the results obtained can at once be expressed in terms of grains per imperial gallon. Since 1 c.c. weighs 1 gramme approximately, therefore 70 c.c., the miniature gallon, weigh 70 grammes or 70,000 milligrammes. Therefore milligrammes per 70 c.c. is equivalent to parts per 70,000 or grains per gallon. In the same way if 100 c.c. is taken for each estimation, each milligramme of constituent found represents directly parts per 100,000, since 100 c.c. = 100,000 milligrammes.

IV. PHYSICAL EXAMINATION.

THE sample is shaken up and poured out into a colourless glass tube 2 feet long and 2 inches in diameter, which has been ground flat at the closed end, and the clearness, presence of suspended matter, colour, etc., observed.

Clearness.—The degrees of turbidity may be expressed as 'faintly opalescent,' 'very slightly turbid,' 'slightly turbid,' 'turbid,' etc. The turbidity or haziness in water is caused by the minute particles of suspended matter, the nature of which is subsequently determined by a microscopical examination. The best quality waters are, of course, brilliant and clear, but these qualities cannot be considered as any evidence of purity, for the most grossly contaminated waters may be both bright and clear.

Colour.—This is best observed by looking down the tube upon a white porcelain slab. Good waters possess but little, if any, colour when thus examined, at the most only

2

a faint grayish-blue or greenish tint. If any yellow or brownish tinge is seen, it will probably be due to the presence of a large amount of sewage pollution, or vegetable contamination in the form of peat, which frequently occurs in upland surface waters. Finely-divided oxide of iron and clay may also give rise to brown tints in water.

A marked greenish tinge may be communicated to water by chlorophyll-containing algæ. Water from rivers that receive the waste waters discharged from various manufactories may be variously coloured, particularly those from dye-works.

A permanent record of the colour of a water can be made by means of the apparatus devised by Messrs. Crookes, Odling, and Tidy. In this apparatus the water is compared against dilute standard solutions of copper sulphate, cobalt chloride and ferric chloride. These are contained in hollow prisms, which are made to slide across one another in front of a circular hole in a metal shield. Thus any desired combination of blue and brown tints can be obtained, and the colour of the water under examination is stated in so many millimetres thick of blue, and so many of brown.

The colour of a water, as well as the degree of turbidity may be determined and recorded quantitatively by the use of Lovibond's tintometer. This ingenious instrument deserves to be better known to analysts generally, and as its price is not prohibitive, would prove of considerable value in water analysis.

The colour of a water is determined by filling a cell fitted with glass ends with the sample water, and having directed the instrument at a white piece of paper a few feet away, the colour is matched by interposing pieces of glass of graduated tints.

The glasses used are red, blue and yellow, varying from a tint so light as to be hardly perceptible up to a bright colour. By means of these standard glasses any tint can be matched, and not only can a measure and analysis of a colour be obtained in terms of so much yellow, blue, green, etc., but the opacity due to suspended matter can be precisely measured.

The instrument requires no special skill on the part of the operator, provided he possesses normal vision and works in average daylight.

Smell.—The odour of a water can be best ascertained by placing about 200 c.c. of the sample in a chemically clean stoppered bottle of about 250 c.c. capacity, and immersing this in warm water, at about 65° C., for a few minutes. The stopper is then removed and deep sniffs taken at its contents. Any sewage or urinous odour, or any contamination with coal-gas, sulphuretted hydrogen, etc., is thus easily detected.

A more delicate method than this is that of M. Boudriment, who shakes 100 c.c. of the suspected water with an equal volume of ether, which is then separated off and allowed to evaporate at the temperature of the room; the residue left after the evaporation of the ether is then carefully smelt. The test may be rendered yet more delicate by distilling a litre and collecting the first 50 c.c. that come over, and treating them with ether as above described.

Many natural waters, *e.g.*, peaty waters, give off peculiar and characteristic odours, which are very difficult to describe. The test of smell is, however, very unreliable, and is one into which a great deal of personal equation enters. Absolutely pure water is quite free from odour, but waters from some geological formations occasionally have a faint odour of sulphuretted hydrogen, particularly those from some sandy and clay formations.

It must not be overlooked that certain 'mineral' waters contain large amounts of sulphuretted hydrogen.

Certain infusoria and algæ may be the cause of bad odours in water supplies.

The *Uroglena Americana* is frequently the source of a ',fishy' odour, and is a serious source of trouble in certain water supplies in the United States. Others give rise to an odour recalling cucumbers. The *Volvox globator* also gives rise to a fishy odour and taste. *Bursaria gastris* gives rise to a seaweed-like odour, and the diatoms *Asterionella* and *Tabellaria* sometimes give rise to an aromatic odour in water-supplies. Recently the Cheltenham public water-supply suddenly acquired a fishy odour and flavour, which was found to be due to a copious growth of *Chara fœtida* upon the walls of the reservoirs.

Taste.—The pleasant taste of a good water is mainly due to the gases, such as carbon dioxide and air, contained therein. The flat taste of rain and distilled water is due to the absence of these gases. The sense of taste is practically useless as an indication of the purity or otherwise of a sample of water. Common salt can be present in amounts as large as 60 grains to the gallon, without causing a brackish taste. Iron salts, however, in very small amounts (as little as 0·2 grains per gallon) will give rise to a very decided chalybeate taste.

As stated above, infusoria and algæ may be the cause of various peculiar and unpleasant flavours in waters.

A certain amount of discretion should always be exercised by the analyst before tasting samples the history and origin of which is unknown, as, for obvious reasons, the safety of such a procedure may be very questionable.

Reaction.—The degree of alkalinity or acidity possessed by a sample of water is not infrequently a factor of great importance. Most natural waters will be found to be faintly alkaline, owing to the presence of calcium bicarbonate, and, less frequently, to sodium carbonate. Many waters are quite neutral in reaction. Badly polluted waters may be alkaline, due to the presence of ammonium carbonate, resulting from the decomposition of the urea contained in urine. Many upland surface waters are decidedly acid, due to humic, ulmic, geic, and other organic acids, derived from peat. In considering the action of water on lead pipes, this reaction is of importance. The waste waters of factories may be excessively acid or alkaline, and these may thus contaminate a large body of water. When necessary, the reaction of a sample of water is determined in terms of centi-normal acid or alkali as the case may be, dilute solutions of methylorange, phenol-phthalein, or lacmoid being used as indicator. If the acidity disappears on boiling, it is due to carbon dioxide. Lacmoid possesses the advantage over phenol-phthalein that it is unaffected by carbon dioxide.

Aeration.—A good water, to be palatable, must be well aerated. The highest-class waters are invariably well aerated. Waters from deep wells, from the chalk and limestone, are frequently highly charged with carbon dioxide. Evidence of this is frequently afforded by minute gas bubbles collecting upon the sides of the bottle containing the sample. The estimation of the dissolved gases in water, which is sometimes required, will be dealt with later.

MICROSCOPIC EXAMINATION OF SUSPENDED MATTERS.

To examine water for suspended matters it was generally the custom to set aside about 200 c.c. in a conical glass, and after the suspended matters had settled out, to examine some of the sediment under the microscope. This plan would doubtless serve in the case of most of the suspended matters likely to be found in water, but as some of them are motile, and others of nearly the same specific gravity as water itself, a more efficient plan is the use of the 'micro-filter,' described by Dibdin (*Analyst*, xxi., 2), by which means the whole of the suspended matter may be rapidly removed from a considerable volume of water in a short time, and a tolerably exact estimation of its quantity arrived at. Dibdin first filters about 500 c.c. of the water to be examined through hardened filter-paper, and then washes the suspended matter retained by the paper into his 'micro-filter,' which is a narrow tube plugged at the smaller end with a mixture of clay and kieselguhr.

Leffmann describes a method modified by Williston from that of Sedgwick and Rafter. In this a small glass funnel has its lower end plugged with absorbent cotton-wool, and a layer 3 or 4 mm. deep of precipitated silica (made by decomposing silicon fluoride with water) is placed upon it. The water under examination is filtered through this, the cotton removed, and the silica with the organisms which it has arrested is washed down with distilled water on to a glass plate. This plate is ruled in fine squares, and forms the bottom of a cell with brass sides holding 2 c.c. of water. The organisms are allowed to settle and then examined.

As ordinary filter-paper is made from linen fibre, one must be prepared to find linen fibres in water that has been examined in this way.

The *mineral matters* that may be found in water are sand, clay, oxide of iron (which sometimes occurs as rounded, semi-transparent masses), etc., and which will settle out on standing, and are of no hygienic importance.

The *organic matters* may be living organisms, animal or vegetable, together with animal and vegetable débris.

For the sake of convenience we divide them as far as possible into classes, giving their significance where it is known. Some of these—for example, desmids and diatoms —are found in the purest mountain water, and are rather evidence of purity than the reverse ; and, speaking generally, the same may be said of most of the chlorophyll-containing organisms. The organisms likely to be found in a water will naturally vary to a large extent with the food available for their sustenance. If the water contains organic matter in solution, it can support the schizomycetes and other organisms which, having no mouths, must take their food in solution. The higher water organisms, which are provided with mouths, or their equivalent, can ingest solid particles. Probably the population of a stream or river is in a state of constant change, as one series of organisms flourishes and becomes extinct, having prepared the way for a fresh race. *These remarks do not apply to sand-filtered water*, which ought to be *entirely free* from all kinds of suspended matter, and the determination of suspended matter, as suggested by Dibdin, is of itself a good means of keeping a check on the proper working of filter-beds.

Algæ.

The algæ generally may be regarded as characteristic of a water poor in organic matter and exposed to the light. They, in company with other organisms containing chlorophyll, are able, with the aid of light, to assimilate carbon from the carbonic acid dissolved in water.

The presence of algæ in water may become important when they are present in large quantities, as they may die and putrefy.

Amphiprora ornata (1) ; *Amphora ovalis* (2).—Both these organisms are diatoms. They contain chlorophyll, and are found in ponds and streams.

Asterionella formosa. — This organism

1.   2. belongs to the diatoms. It contains oils or fats in its ectoplasm, and produces a disagreeable, fishy taste and smell in water. It is a rod-shaped body, somewhat dumb-bell shaped.

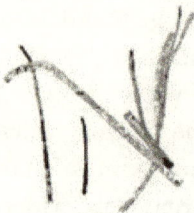

Beggiatoa alba (3).—This organism is characteristic of decaying vegetable or animal matter. It is distinguished by the bright grains (said to be sulphur) visible in the cells. It is common in sulphur springs, and so prevalent in polluted water as to have earned the title of

3. luted water as to have earned the title of the sewage fungus. It contains no chlorophyll.

Cœlosphærium (4).—This organism is mentioned by E. Waller (*Board of Health Reports, Brooklyn*), as producing a disagreeable smell in water.

4.

Cladothrix dichotoma (5).—This alga is found more frequently in stagnant water than in running streams or good well-water. The filaments are harsh to the touch. It is common in the refuse-waters from sugar-factories. In Russia it often occurs in town-water supplies. It grows readily on the surface of putrefying animal or vegetable matter immersed in water. It has an avidity for iron, and sometimes forms obstructions in iron pipes.

The cladothrices are distinguished from the crenothrices by their 'false branching.'

5.

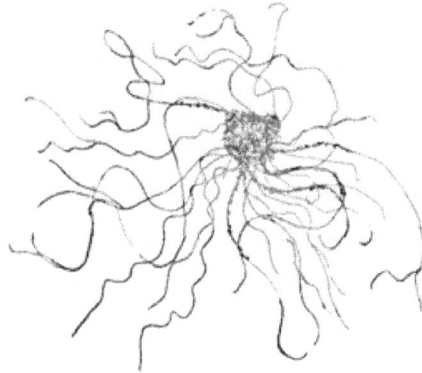

6.

Cladothrix dichotoma (6) under a low power.

Conferræ (7).—These are filiform algæ, which are widely distributed in all ponds, rivers, and reservoirs, being especially abundant near the margins. They are characteristic of water comparatively poor in organic matter.

7.

Crenothrix (or *Leptothrix*) *Kühniana* (8).—This organism is common in running or stagnant water that is rich in organic matter. The threads are motile, and in their younger stages are grouped into little patches. They have a greenish or brownish colour, and impart the same tint to the water. If a reservoir is infected with them, large quantities of water may be deteriorated in a short time. The luxuriant growth of this organism at the Berlin water-works proved so troublesome that fresh filter-beds

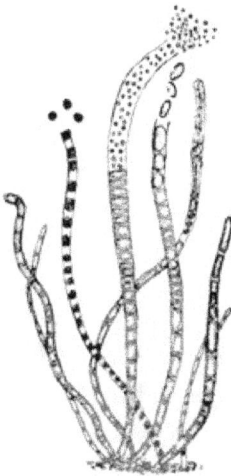

8.

were constructed. It developed to a thickness of some feet

in the streams supplying the water-works, although the organic matter dissolved in the water was small.

9.   10.

Crenothrix (9, 10) under a low power.

Closterium ensis (11); *Closterium dianœ* (12); *Closterium rostratum* (13).—These organisms are among the commonest of the desmids, and are often found in water-

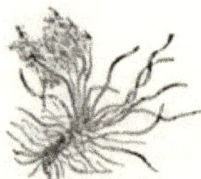

11.   12.   13.

supplies. Their presence is not objectionable, as they can only live in pure water which has plenty of air and sunlight. They all contain chlorophyll.

14.   15.   16.

Cosmarium (14) contains chlorophyll, lives in open standing water, and is, therefore, not often found in the examination of drinking-water.

Gomphonema acuminatum (15); *Gomphonema capitatum* (16).—Occur chiefly in pond-water. They contain chlorophyll, and belong to the diatomaceæ.

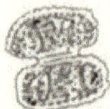

17.

Melosira granulata (17).—This diatom forms long filaments in water, and, on account of oils or fats in its ectoplasm, produces a disagreeable grassy taste and smell in water-supplies.

Navicula radiosa (18); *Navicula viridis* (19).—These diatoms are motile, and contain chlorophyll of a dark yellow colour.

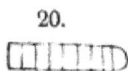

20.

21.

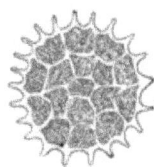

18.　　　　19.　　　　21.　　　　22.

Oscillaria (20) ; *Nostoc* (21).—These organisms, which are anabœna, form bluish-green or brownish patches. They are chiefly found in stagnant water ; and as they produce a disagreeable smell in water, their presence is very objectionable on this account, and because they indicate the presence of a large amount of organic matter.

Pediastrum (22).—It contains chlorophyll, and is found in pond-water.

Protococcus viridis (23).—This is a very common inhabitant of streams and rivers, and probably assists in their self-purification. It propagates by fission, and is generally seen in groups of four.

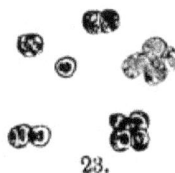

23.

Scenedesmus (24).—This organism is common in ponds, but does not appear to possess any hygienic importance; it contains chlorophyll.

24.

Spirogyra (25). — This organism contains chlorophyll, in spiral bands, and starch granules. It is a common inhabitant of ponds and streams.

25.

Staurastrum gracile (26).—This organism is one of the commonest desmids, and its presence in water is not objectionable—in fact, it is stated that it can only live in pure water which is well aerated and exposed to the light.

Stauroneus phœnicenteron (27).—This organism is one of the diatoms, and contains chlorophyll.

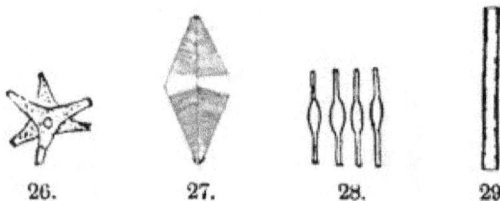

26.   27.   28.   29

Synedra pulchella (28) ; *Synedra ulna* (29).—Both these diatoms are said to cause a disagreeable grassy taste and smell in water.

Fungi.

The moulds, *Mucor mucedo* (30), *Aspergillus* (31), and *Penicillium glaucum*, are not to be regarded as natural inhabitants of water, since they do not multiply therein.

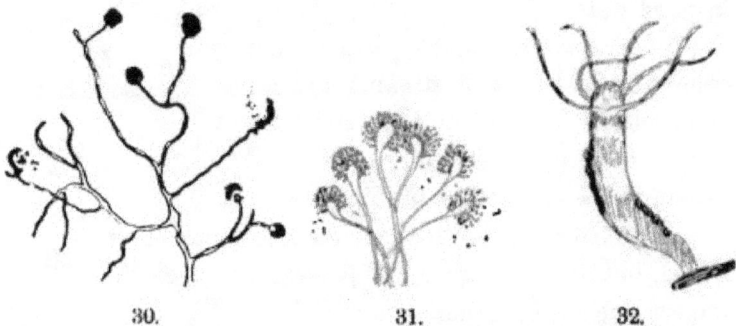

30.   31.   32.

Their natural habitat is moist decaying organic matter, and therefore their presence in water is objectionable.

Cœlenterata.

Hydra vulgaris (32).—This organism is characteristic of stagnant water.

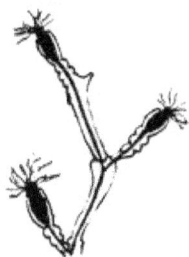
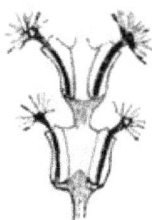
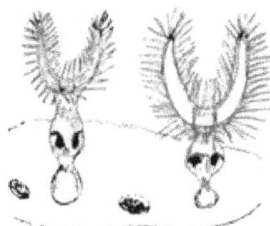

33.   34.   35.

Polyps, Sertularia (33) and *Campanularia* (34); *Polyzoa, Lophopus* (35), *Cristatella* (36).—These are low animals allied to the hydra, corals, and anemones.

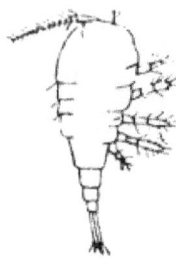

36.   37.   38.

Crustaceæ.

Bosmina longirostris (37); *Cyclops* (38); *Daphnia, water flea* (39).—All these organisms are characteristic of pond-water, and should not be present in a water-supply. The crustaceans feed on the algæ, on decomposing vegetable matter, and on the smaller ciliata. One or two specimens may be found in a good

39.

water, but if many are present they indicate the presence of large quantities of decomposing organic matter.

Infusoria.

It may be said generally of the infusoria that they are characteristic of pond-water containing suspended dead organic matter. Most of the flagellated infusorians are to be found in putrid vegetable infusions in immense numbers, and they ought not to be present in pure water, except perhaps very sparingly.

Anthophysa (40).

Ceratium (41).—A fairly large infusorian. It is goblet-shaped, and has long projections or spines round the mouth part. The opposite end is drawn out into a spine.

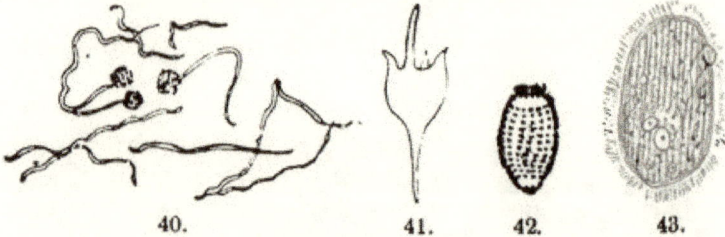

40.   41.   42.   43.

Coleps (42) is a globular infusorian, looking like a basket, water having access through the meshes or pores of the body. It is covered with cilia, and has an anterior and a posterior opening, both surrounded by short spines.

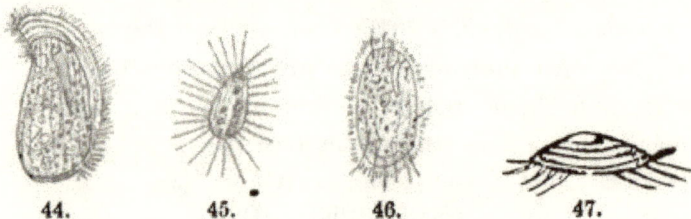

44.   45.   46.   47.

Colpidium (43).   Characteristic of dirty water.

Chilodon (44).

Cyclidium (45).

Euplotes Charon (46).   In a different position (47). Characteristic of dirty water.

Euglena viridis (48).—It is a green animal containing chlorophyll, and lives wholly on vegetable foods. It has a

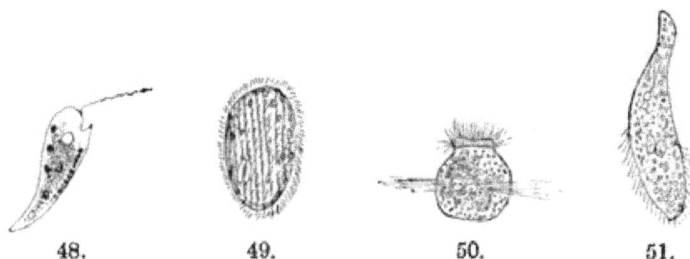

48.   49.   50.   51.

long flagellum, at the base of which is a gullet. It has a contractile vacuole, and is actively motile. There are two other organisms similar to the *Euglena viridis*, namely, the *Euglena oxyuris*, which is large, and *Euglena spirogyra*, which is intermediate in size between the other two.

Glaucoma (49).

Halteria (50).

Oxytricha (51).

Phacus longicaudis (52) is a heart-shaped infusorian. At the anterior end is a single long flagellum. The posterior end is drawn out into a long spine.

52.

53.

Paramœcium bursaria (53). — Commonly called the 'slipper animalcule.' It is covered with cilia, and has a flattened form. The mouth is on the ventral side, a little to one side, and leads into a gullet. It has a layer of trichocysts below the ectoplasm (skin). These are pointed rods, which it ejects when irritated. It has two nuclei.

There appear to be many different varieties of paramœcium, one of which at least has pathogenic properties ; and the *Paramœcium coli* (or *Balantidium coli*) is spoken of by Besson as the cause of dysentery in man, though the evidence is inconclusive. Klein regards it, however, as diagnostic of sewage pollution.

Paramœcium putrinum (54). This organism is nearly always present in putrid vegetable infusions.

54.

Paramœcium coli (Balantidium coli) (55).

Polytoma (56).—It is the simplest of the infusoria. It is

 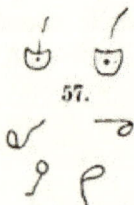  

57.

55. 56. 58. 59. 60.

saphrophytic, and takes its food in solution, living in putrefying animal or vegetable matter. It has no mouth, and absorbs its food by the osmotic power of the body substance. It has two flagella, and there is a delicate invisible cuticle bounding the body.

Paramonas globosa (57).—Very small, heart-shaped, with one flagellum.

Pleuromonas jaculans (58).—Small pear-shaped organisms with one flagellum. Certain organisms very similar in size and appearance are described by Besson as occurring in the fæces in diarrhœa. He names them *Cercomonas intestinalis* and *Cercomonas termo ;* the latter is found in putrid vegetable infusions. It is classed among the protozoa, and is possessed of active motility.

Stentor (59) is trumpet-shaped, and fixed by the narrow end. The body is uniformly covered with cilia, and at the front end of the body is a spiral layer of longer cilia. This organism is characteristic of stagnant water.

Stylonichia (60). — It is shaped something like paramœcium. It has a strong layer of cilia round the mouth. The dorsal surface has no cilia, but has stiff hairs, which are really immovable cilia.

Synura uvella (61).—This forms round, irregular clumps.

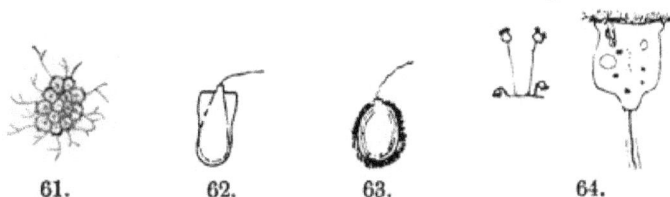

61.   62.   63.   64.

Each individual is round and ciliated, and has one long bifid flagellum.

Trachelomonas cylindrica (62), *piscatoris* (63).—These organisms are bottle-shaped. At the anterior end is a single large flagellum set in a collar.

Vorticella (64). — This organism is cup-shaped, and is attached to the substratum by a long contractile stalk. The body has a lid, and both the lid and the opening into the gullet are strongly ciliated. The nucleus is horseshoe-shaped. It is characteristic of stagnant water.

Volvox globator (65).—A green, globular colony of individuals. The colony is spherical, and the individual creatures live in the wall of the sphere. Each has two flagella and a nucleus, and is of a green colour. The ball, or hollow colony, continually rotates by means of the flagella of each individual. This

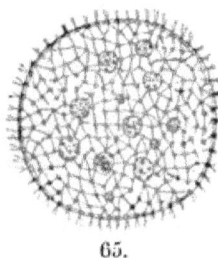

65.

organism gives rise to a fishy taste and smell in water.

3

Insecta.

Macrobiotus ursellus (66). — Commonly known as the

66.

'water-bear.' It has four legs, ending in stiff cilia, which help it to hold on to water-plants. The head has no antennæ, and has eyes ; the mouth is suctional. It is microscopic in size, and is found in pond-water.

Porifera.

Spongilla fluviatilis.—The only fresh-water sponge. It grows in greenish masses in canals and fresh-water lakes. The sponge itself is an inch or more in diameter; the spicules are most likely to be noticed.

Protozoa.

Actinophrys sol (67).—This is a small round animalcule, with one nucleus. It is surrounded by stiff, radiating pseudopodia. Found in pond-water.

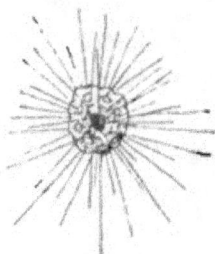 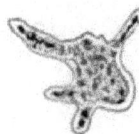 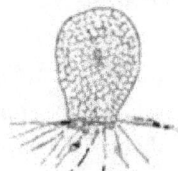

67.   68.   69.

An organism called *Actinospherium Eichornii* is very similar in appearance to the *Actinophrys sol*, but much larger.

Amœba (68).—Besson describes three varieties, namely, *Amœba princeps*, *Amœba coli*, and *Amœba jelaginia*. The *Amœba coli* may be pathogenic, and appears to be distinguished by only extruding one pseudopodium. It is one of the simplest animals, and consists of a shapeless mass of protoplasm, which is constantly flowing out into blunt, uneven, lobose prolongations—the pseudopodia. It has no mouth, and the food is taken right into the protoplasm and digested, and the undigested part thrown out again.

Difflugia (69).—This organism has a coating formed by sand which it has taken into its body, and with which it makes a globular shell, inside of which is the soft protoplasm of the body-substance. This flows out at the mouth of the shell, and forms anastomosing threads, which catch the food of the animal, acting as a net.

Arcella (70).—This organism has a shell, which partly covers it. On the flat ventral side is an aperture, through which the naked protoplasm exudes as lobose pseudopodia. In the body-substance are seen vacuoles containing gas. These are hydrostatic in function.

70.

Rotiferæ.

The rotiferæ generally are characteristic of stagnant water full of organic matter, and when found in large numbers should cause the water to be regarded with grave suspicion.

71.   72.   73.   74.

Anurea cochlearis (71).

Polyarthra platyptera (72).

Rotifer vulgaris (73).—The first to be discovered. Gives

a red or green colour to gutter-water. It is often called the 'wheel' animalcule, on account of a circular oral disc, which is fringed with cilia, which is motile, and by its movements conveys an appearance of rotation, and serves to propel the animal and to set up food-currents.

Vermes.

Anguillula fluriatilis (or *aquatica*) (74).—These organisms are very probably the young forms of intestinal parasites, and their presence in water is highly objectionable.

Ova of Parasites affecting Man.

Tænia solium (tape-worm) (75).
Bothriocephalus latus (76).
(77) The same in a more advanced stage.
Ascaris lumbricoides (round worm) (78).

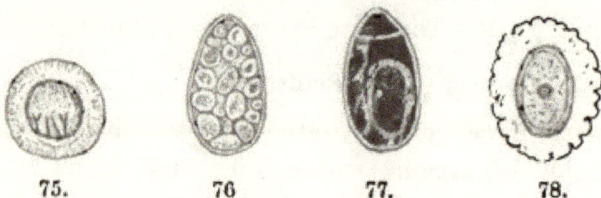

75.   76.   77.   78.

Oxyuris vermicularis (thread-worm) (79).
Anchylostomum duodenale (80).
Trichocephalus dispar (81).

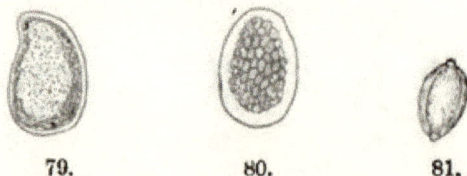

79.   80.   81.

Newt (*Salamander*).—The common newt may become parasitic to man through ingestion of the eggs, and as the consequences are very grave, any water containing these amphibia should be absolutely prohibited as drinking-water.

Animal and Vegetable Débris.

Fibres of clothing, such as linen (82), wool (83), silk (84), and cotton (85), may be found in polluted waters.

Epithelial scales (86).

Undigested muscular fibres (87).

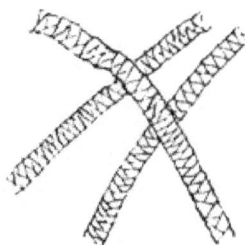

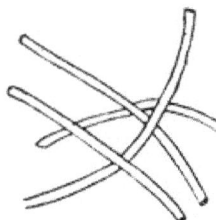

82.   83.   84

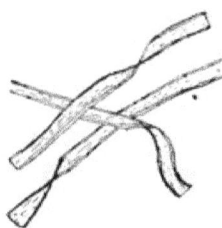

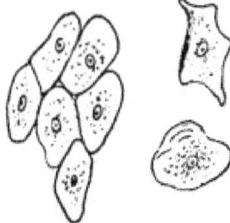

85.   86.

87.

V. CHEMICAL EXAMINATION.

Total Solids.—The amount of solid matter left on the evaporation of a given quantity of water will naturally vary largely according to the geological character of the locality from which it has come. The 'total solids' of a potable water consist mainly of calcium carbonate, sodium chloride, small quantities of nitrate and sulphate of calcium, and a little silica, together with any non-volatile organic matters that may be present.

If it is not desired to ascertain what the total solids consist of, but merely to learn their amount, it will be sufficient to work on 50 or 100 c.c., which should be evaporated in a platinum dish. The dish must be cleaned with very fine pumice powder till quite bright outside and in ; and the evaporation may be begun over a low gas-flame, but must be finished over a water-bath.

As a temperature of 100° C. is not sufficient to expel all moisture, the dish should then be placed in an air-bath and heated up to about 105° C. for at least twenty minutes. It is then placed in a desiccator till just cold, and weighed without delay. Not only do the total solids absorb moisture with great readiness, but the dish itself becomes heavier, owing to the property possessed by platinum of occluding gases on its surface, so that a dish weighing 25 to 30 grammes may gain 2 milligrammes on standing for half an hour after heating. It is therefore essential that the dish should be weighed within a definite time, say four minutes, after placing in the desiccator, whether it contain total solids or be weighed empty.

After the weight of the total solids has been obtained, it is well to note carefully their behaviour when they are exposed to a very low red heat. If an appreciable quantity of organic matter is present when heat is applied, the solids will darken, and, on smelling, an odour of weeds or of burning animal matter may be perceived, according to whether vegetable or animal pollution is present, and in most polluted waters the smell produced is very distinctive. The temperature at which the ignition should be done is the lowest possible red heat, which must be continued just long enough to cause any darkening to give place to a white or gray, thus showing the complete combustion of the organic matter. By heating in this way we shall obviously expel carbonic acid from any carbonate of calcium that may be in the residue ; and if the weight of the solids after ignition is subtracted from that of the total solids before ignition, the loss will be due to organic matter and to expelled carbonic acid. As it is our object to obtain as closely as possible the true weight of combustible matter, we replace the expelled carbonic acid by moistening every part of the residue with a strong solution of ammonium carbonate. After drying, the dish is very cautiously heated so as to expel all excess of ammonium carbonate, but not to causticize the calcium carbonate afresh. Although it is more than probable that the figures we obtain in this way are liable to considerable errors, the estimation is well worth making, for it will rarely happen that a good water, except possibly a peaty one, will yield a residue that will either darken or smell much on ignition, while there are very few bad waters that will not betray their character by the one or other indication. The 'loss on ignition' in a water of good quality will seldom amount to more than 20 per cent. of the total solids. Generally speaking, the total solids should not exceed 40 or 50 parts per 100,000 ; but, of course, much depends upon the constitution of the solids and upon the geological source of the water. Except in the case of waters from peaty sources, the slightest evidence of fumes, or darkening of the solids on ignition, will point to organic contamination.

According to Drown and Hazen, it is possible to completely incinerate all the organic matter without expelling CO_2 from $CaCO_3$. These analysts, following Young and others, add a known quantity of sodium carbonate to the solution before evaporation, in this way precipitating lime and magnesia as carbonates, and providing acid radicles with an alkaline base. The solid residue does not contain water of crystallization. It is incinerated in a 'radiator' made of a platinum dish, which surrounds the dish in which the evaporation has been conducted, and a platinum dish is placed over the latter, round which an air-space of about half an inch is left. The radiator is brought to a bright red heat, which is stated to burn off all organic matter without decomposing calcium carbonate or sodium or potassium nitrate, and without volatilizing alkaline chlorides. In practice a loss does occur, especially with magnesium salts ; but the process is doubtless more accurate than the ordinary determination.

Analysis of the Mineral Residue.—The determination of the constituents of the mineral residue in a water is of little practical importance in the sanitary examination of water where the mineral residue is small in amount ; but where this is large, and in cases where the water is employed for the raising of steam in boilers, the determination of the nature of the inorganic constituents is a matter of great importance. Much information is gained in this connection by taking into consideration the ratio of permanent to temporary hardness (*q.v.*) to the mineral solids.

The full analysis of the mineral constituents is a very lengthy and tedious operation, and is seldom required in practice, except in the case of natural mineral waters. A full account of the best methods for this operation will be found in Fresenius's ' Quantitative Analysis.'

The following scheme for the partial mineral analysis will be found sufficient for most purposes.

One thousand c.c. of the sample is acidified with HCl and evaporated in a platinum dish by first gently boiling down to a small bulk over a naked flame, finishing on a water-bath, and drying to constant weight in a hot-air oven. After weighing the total solid residue, it is ignited, and the residue recarbonated with a few drops of ammonium carbonate solution, dried, and gently ignited until ammoniacal fumes cease to be evolved. The residue will be the fixed mineral matter, and the loss on ignition will equal the organic and volatile matter.

Silica.—The mineral residue is treated with a little HCl and evaporated to dryness, which operation is once repeated. The residue is again acidified with hydrochloric acid, warmed with water and filtered. The residue is silica, SiO_2, which is washed and weighed. In some cases the presence of silica in a water is of great importance, as its presence diminishes the plumbo-solvency of waters liable to be contaminated with lead.

Iron and Aluminium.—The filtrate from the silica determination is acidified with a few drops of strong nitric acid free from iron, and boiled, so as to convert any iron salts into ferric. A little ammonium chloride solution and a slight excess of ammonium hydrate are then added, and the precipitated hydroxides filtered off, washed, dried, and ignited. The iron and aluminium are weighed, and calculated as $Al_2O_3 + Fe_2O_3$.

An alternative estimation of iron alone may be made by

acidifying with a few drops of iron-free nitric acid, evaporating
to dryness, dissolving the residue in 50 c.c. of water, adding
1 c.c. of very dilute potassium ferrocyanide solution, and
comparing in a Nessler glass with 50 c.c. of a ferric solution
of known strength to which the same quantity of the
potassium ferrocyanide solution has been added. If iron
be present in the water, a blue colour will develop on standing; and by pouring off from the Nessler glass, which shows
the darker tint, until the columns of liquid in the two
glasses, placed on a white tile, are of the same depth of tint,
the quantity of iron present in the water can be approximately calculated. A suitable standard iron solution is
prepared by dissolving ammonium ferrous sulphate in
recently - distilled water, boiling with nitric acid, and
diluting to the strength of ·7 gr. of salt (= ·1 gr. Fe) per
litre.

Calcium.—NH_4HO and $(NH_4)_2C_2O_4$ are added to the
filtrate from the iron determination, the beaker being
allowed to stand upon a hot-plate for some time, after
which the calcium oxalate is filtered off, washed, and
ignited, and the calcium is weighed as CaO.

Magnesium.—The filtrate from the above is treated with
Na_2HPO_4, well stirred, and allowed to stand for a few
hours. The precipitate is filtered off, washed with 8 per
cent. solution of NH_4HO, and ignited; the magnesium is
then weighed as $Mg_2P_2O_7$. Factor to convert $Mg_2P_2O_7$ into
MgO = ·36216.

Alkalies.—When these are present, they are determined
as follows : 1,000 c.c. of the sample is evaporated to a small
bulk, cooled, and filtered ; the filter is well washed with
boiled and cooled distilled water. The filtrate is then
titrated with $\frac{N}{10}$ H_2SO_4, using methyl-orange as indicator.
The result is calculated as Na_2CO_3. One c.c. $\frac{N}{10}$ $H_2SO_4 =$
·0053 Na_2CO_3.

Sulphates.—These are determined upon 100 c.c. of the sample by acidifying with HCl, and precipitating with $BaCl_2$. The precipitated $BaSO_4$ is filtered off, ignited, and weighed, and the result calculated to (SO_3). Factor $= \cdot 41216$.

Chlorides.—These are determined by the titration of 100 c.c. of the sample with a standard solution of silver nitrate by the method described on p. 48.

Statement of Results.—The above approximate scheme of the analysis of a water residue will be found to yield all the information that is generally required. If it is desired to express the results in combination, it is necessary to bear in mind their respective chemical affinities; that is to say, the strongest acid will combine with the strongest base, due regard being paid to the relative solubility of the salts formed. It may be assumed that the whole of the chlorine is in combination with sodium, but if there is excess of sodium, it is combined with sulphuric acid. Any excess of sulphuric acid is combined with calcium and magnesium. Calcium and magnesium present not in combination, as sulphates or chlorides, may be assumed to be in the form of bicarbonates, which may be precipitated on boiling.

Hardness of Water.

The hardness of water is generally due to the presence of salts of calcium and magnesium. That which is due to either carbonate is known as temporary hardness, owing to the fact that it can be destroyed by boiling. When such a water is boiled, these carbonates are precipitated, owing to the excess of carbonic acid gas which holds the carbonate in solution being driven off. This is the cause of the deposit or 'fur' in kettles and 'scale' in steam-boilers. The hardness which cannot be destroyed by boiling is known as permanent hardness, and is due to sulphates of the alkaline earths.

The hardness of water is most important both from the economic and hygienic standpoint. Hard waters, particularly those having excessive permanent hardness, may give rise to gastric and intestinal trouble in some individuals. Moreover, the use of hard waters entails a great waste of soap in laundry work and other trade processes, and great loss of heat results in boilers used for steam-raising. It is estimated that $\frac{1}{4}$ inch of incrustation in a boiler involves a loss of over 40 per cent. of the energy of the coal used.

Estimation of Total Hardness.—The hardness of a water is readily estimated by ascertaining how much of a standard soap solution is required to produce a lather lasting for two minutes with a certain volume of the water.

In making the determination, a standard soap solution is run into a solution of a calcium salt till a lather is formed; each equivalent of lime destroys an equivalent of soap. Put in an equational form in the case of calcium sulphate and sodium oleate,

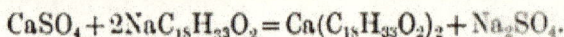

$$CaSO_4 + 2NaC_{18}H_{33}O_2 = Ca(C_{18}H_{33}O_2)_2 + Na_2SO_4.$$

A convenient quantity of water to take for the purpose is 100 c.c., and then, if our soap solution contains just sufficient soap per cubic centimetre to be exactly precipitated by a milligramme of chalk, it is clear that the number of cubic centimetres of soap precipitated by 100 c.c. of the sample of water will indicate the number of parts per 100,000 of chalk in the water, after allowance has been made for the amount of soap needed to produce a lather in pure water. It is clear that a certain amount of soap solution will be required to make a lather even in a water containing no hardness at all; and if the experiment be tried, using 100 c.c. of distilled water, it will approximately be found that 1 c.c. of soap solution makes a lather lasting about two minutes. It is therefore customary to subtract 1 c.c. from the number of cubic centimetres of soap solution required to make a lather with the sample water.

Preparation of Standard Soap Solution.—Soap solution is most easily prepared by dissolving 10 grammes of white Castile soap in fine shavings in methylated spirit. After warming, the soap will dissolve, but the solution will probably be cloudy, and must be filtered through filter-paper. This solution has now to be tested against a water of known hardness, and if too strong, some more spirit must be added, or if too weak, some more soap. The soap solution will not keep at the same strength, but gradually becomes weaker, and it is therefore best to test it every time before it is used. For this purpose we require a solution of standard hardness, which is generally made by dissolving a gramme of Iceland spar—a pure form of calcium carbonate—in hydrochloric acid, evaporating to expel excess of hydrochloric acid, and making up to a litre with distilled water. Such a solution will then contain the equivalent of 1 milligramme of calcium carbonate in every cubic centimetre. We are obliged to dissolve the Iceland spar in an acid because neither Iceland spar nor any other form of calcium carbonate is soluble to the extent of more than three parts per 100,000. We have therefore to assume that the soap-destroying power of the calcium is the same when dissolved as calcium chloride as it is in an ordinary drinking-water, where it exists mainly in the form of calcium carbonate kept in solution by excess of carbonic acid gas. Supposing, then, that we are testing a fresh batch of soap solution, we should place the soap solution in a burette, and take 10 c.c. of the solution of standard hardness, make it up to 100 c.c., and run in the soap, as in testing an ordinary water.

If the soap solution is just of the right strength, a lather, lasting two minutes, should be produced by 11 c.c.

In making tests, a stoppered bottle holding about 150 c.c. should be used, and the soap solution is added at about the rate of 2 c.c. at each successive addition; when the liquid shows signs of lathering, the bottle should be laid on its side, and the lather should last, as nearly as possible, two minutes. Great care should be taken not to mistake the peculiar scum yielded by magnesium salts for the typical soap lather.

In the case of hard waters, in which the two-minute lather is not yielded by 15 c.c. of soap solution, the water must be suitably diluted. In presence of substantial amounts of magnesia various factors have been stated by various analysts as representing the correction to be applied to the results. Probably the best correction is to disregard them altogether, and determine the calcium and magnesium separately.

Estimation of Permanent Hardness.—This is estimated by boiling 100 c.c. of the sample down to about 40 c.c., when the chalk will be precipitated; the liquid is then filtered off and made up to 100 c.c. with cooled boiled distilled water, and tested with the soap solution, as mentioned above. Each c.c. of soap solution represents a hardness equivalent to 1 milligramme per 100 c.c. of $CaCO_3$, or 1 part per 100,000 of 'permanent hardness.' This figure, deducted from that representing 'total hardness,' will give the 'temporary hardness.'

Hehner's Method for the Estimation of Hardness.—In this process the calcium and magnesium carbonates, which cause the temporary hardness, are first determined by titration with standard acid. Both the carbonates and sulphates of calcium and magnesium are then precipitated by adding a known volume of standard sodium carbonate solution in excess, and boiling for half an hour. The unchanged sodium carbonate is then estimated by means of standard acid. An equal quantity of untreated water should be similarly boiled, and the amount of sodium carbonate which it contains estimated for deduction from the figure given by the process. The amount of sodium carbonate which is decomposed by the above sulphates is thus found, and from this the permanent hardness can be calculated.

Temporary Hardness.—70 c.c. of the sample are made hot in a dish, and then titrated with $\frac{N}{50}$ H_2SO_4, using phenoacetoline (0·2 per cent.) solution as indicator. Each cubic centimetre of acid used is equivalent to 1 grain per gallon of $CaCO_3$.

Permanent Hardness.—Take 70 c.c. of sample, and add about as many cubic centimetres of $\frac{N}{50}$ Na_2CO_3 solution as there are grains per gallon of non-volatile solids. Transfer to a platinum dish, and evaporate to dryness on a waterbath. Take up with recently boiled distilled water, which must be quite cold, then filter, wash the residue well, and make filtrate up to about 70 c.c. This is now raised to boiling-point, and titrated with $\frac{N}{50}$ H_2SO_4, using phenoacetoline as indicator. The number of cubic centimetres of acid used deducted from number of cubic centimetres of $\frac{N}{50}$ Na_2CO_3 used equals grains per gallon of permanent hardness.

The total hardness is obtained by adding the sum of the permanent and temporary hardness.

The results of the determination of hardness are frequently returned in terms of 'degrees.' Each degree of hardness (Clark) corresponds to an equivalent of 1 grain of $CaCO_3$ per gallon. On the Continent, however, 1° of hardness is equivalent to 1 milligramme per 100 c.c., or 1 part per 100,000.

It is of interest to note that each degree of hardness represents a soap-destroying power equivalent to 2½ oz. of soap per 100 gallons of water.

A greater hardness than 30 parts per 100,000 is undesirable from the hygienic standpoint. The higher the ratio of 'temporary' to 'permanent' hardness, the better. The hardness of waters may be classed as follows :

Very soft	...	...	...	3 to 5 parts per 100,000
Moderate	...	...	...	5 to 10 ,, ,,
Hard	...	...	...	10 to 30 ,, ,,
Very hard	...	...	...	above 30 ,, ,,

For successful brewing purposes waters containing large amounts of 'permanent' hardness are valued. In the brewing of light ales the sulphate of calcium tends to keep out the colouring matters of the malt. The water of Burton-on-Trent contains from 28° to 30° of permanent hardness.

Chlorides. — Sodium chloride is a very constant constituent of urine, in which it is present to the extent of about 0·75 per cent., and hence is always to be found in waters that have suffered contamination from sewage or cesspool leakages.

The absence of chlorides in a water may be taken as good evidence of absence of sewage pollution ; but the converse is of course not the case, as chlorides in water are also derived from the geological strata through which the water passes.

As chlorides are not readily removable from drinking-water or sewage by any natural or artificial process, except perhaps by being drawn up by the roots of plants, it follows that when a water has suffered contamination with sewage matters the chlorides will remain as evidence, even after the organic matter that accompanied them has undergone natural purification and has disappeared.

Chlorine is usually estimated by titrating 100 c.c. of the sample, filtered if necessary with a standard solution of nitrate of silver. Decinormal solution of silver nitrate being rather too strong for convenience, it is customary to use a solution, such that every cubic centimetre of silver solution used will indicate ·001 of a gramme of chlorine. Consequently working on 100 c.c. of water, each c.c. of silver solution used indicates one part of chlorine per 100,000.

According to the equation,

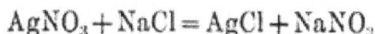

$$AgNO_3 + NaCl = AgCl + NaNO_3$$

(the molecular weight of silver nitrate being 170 and of chlorine 35·5), one part of chlorine will be exactly precipitated by $\frac{170}{35\cdot5}$ or 4·79 parts of silver nitrate, so that a solution, 1 c.c. of which exactly precipitates a milligramme of chlorine, must contain 4·79 milligrammes of silver nitrate per c.c. or 4·79 grammes per litre.

In preparing the solution, some nitrate of silver should be heated in the air-bath nearly up to the fusing-point, and then 4·79 grammes, cooled in the desiccator, weighed out quickly and dissolved in a litre of distilled water. The solution should be kept in a brown-coloured bottle or in the dark. The titration is performed in a white porcelain dish, and a drop of solution of potassium chromate is used as indicator. As soon as all the chlorides in the water have been converted into the silver salt, any further addition of silver will cause the formation of silver chromate, which is brick-red in colour, so that as soon as a perceptible colour is observable the titration is stopped, and the amount of silver solution used up is noted. As chlorides are usually

4

present in potassium chromate, the salt should be purified before use.

It is usual to deduct ·2 c.c. for the amount of silver solution required to change the colour of the indicator.

W. G. Young has shown (*Analyst*, xviii., 125) that titrations to estimate chlorine should be performed always on cool solutions, as an increase of temperature even to 38° C. (blood heat) gives results too high in consequence of the increased solubility of silver chromate with rise of temperature. He also shows that it is preferable to concentrate the solution to be titrated, and not to work with too dilute a solution of silver nitrate.

The amount of chlorine found in a water depends largely upon the nature of the soil through which the water has permeated. Surface-waters and rain-water collected in the neighbourhood of the sea, even at considerable distances from it, frequently contain very large quantities of salt, and in these cases the amount of chlorine found has but very little significance; but in cases where the geological strata contain but little chlorides, and where there are no manufacturing effluents to cause pollution, the amount of chlorides found is a very valuable guide to the approximate amount of sewage contamination present in the water. In the case of waters containing chlorine derived from sea-water, magnesium salts will also be found. In several of the American States, particularly **Massachusetts**, the normal distribution of chlorine has been mapped out. This has been done by estimating the amount of chlorine in the unpolluted waters of the districts at a large number of points. When these points are joined by means of lines, the normal distribution of chlorine at a particular point is easily found. These lines are known as 'isochlors.' From the chart so prepared of the isochlors in Massachusetts, the normal chlorine is found to vary from

0·45 grain per gallon near the coast to less than 0·06 in the western part of the State (*Board of Health Report,* 1892). These plans of the isochlors are exceedingly valuable, as when in a given spot an amount of chlorine is found in excess of the normal it points to practically positive evidence of sewage pollution. It has been estimated by Thresh that every 100 persons per square mile add on the average 0·03 grain of chlorine per gallon to the water flowing from the area in consideration. From this data it is easy to calculate the approximate extent of the pollution. Unfortunately the plan of determination of the isochlors of districts, which is being so systematized in the United States, has but little application in small insular countries like England, where the geological structure is not continuous over large areas, although much valuable information might be gained by the systematic determination of the normal chlorine in the unpolluted waters of given areas. It should be noted that the value of the chlorine determination is largely detracted from by the fact that the organic matter which originally accompanied the pollution may have become completely oxidized. Therefore it should be noted that the chlorine figure does not necessarily denote present pollution, although it may to a great extent be a measure of the past pollution, if it exceeds the normal chlorine figure of the district, which is due to geological conditions.

Generally speaking, unless it is accounted for by the geological conditions of the district, a greater amount than 2 parts per 100,000 should be regarded with suspicion.

The peculiar value of the chlorine determination lies in the stability of chlorides in solution, and the readiness, therefore, with which in a series of examinations, such as are indispensable for a sound opinion as to the safety of a supply, any sewage pollution can be detected by its increase.

Phosphates.—Phosphates in water can in most cases be

only of sewage origin, or derived from organic matters such as occur in the neighbourhood of graveyards. Absence of phosphates, however, is no proof that the water is free from sewage pollution, since substances like iron and alumina, which are present in all soils, precipitate phosphoric acid ; and plants derive their phosphorus from it, and thus remove it from the soil.

As small a quantity as ·0005 gramme of calcium phosphate will yield a yellowish turbidity with ammonium nitrate and ammonium molybdate, so that any appreciable quantity would be detected by treating the ignited total solids from 250 c.c. of water with a little nitric acid, neutralizing with ammonia, evaporating the filtered liquid in a porcelain dish to a bulk of ·5 centimetre, adding a drop of strong nitric acid and a few drops of 10 per cent. solution of ammonium molybdate, and warming the whole to 70° C. If it is desired to estimate the amount present, 2000 c.c. of the water should be evaporated and the same procedure carried out, the precipitate collected on a small filter, washed with alcohol and ether and dried, when the weight obtained, multiplied by ·0373, will give the result as P_2O_5.

O. Hehner (*Analyst*, v. 135) published determinations of the phosphoric acid in a number of different waters. He concludes that when the phosphoric acid is above ·5 parts per million the water should be regarded with suspicion, but that the absence of phosphoric acid is not a positive proof of freedom from pollution, as some waters known to be polluted when examined by him contained very little or none.

J. West-Knights (*Analyst*, v. 180) has described a colorimetric process for the estimation of P_2O_5 in waters, which he verified by comparison with the gravimetric process.

Sidney Harvey, in the same number of the *Analyst*, gives results obtained on an old well, which had been closed for some time, situated near an ancient churchyard. He found as much as 5 grains of nitrates and 1 grain of phosphoric acid per gallon.

Poisonous Metals.—The metals which we have occasionally to seek for in water are lead and copper.

These are usually tested for by placing 100 c.c. of the water to be tested in a Nessler glass, and passing a slow current of sulphuretted hydrogen through it, having previously added a single drop of strong hydrochloric acid. If a dark cloud forms it must be due to lead or copper, and we may either proceed by concentrating the water, or we may at once test for lead by adding potassium bichromate or potassium iodide. In both cases a yellow cloud will form after an interval. If lead is present, about $\frac{1}{50}$ grain per gallon (·28 parts per 100,000) can be detected in this manner. An approximate estimation of lead in water can be made by comparing the tint produced on passing sulphuretted hydrogen gas into 100 c.c. of the sample, and then making up a tint to correspond with it by adding so many centimetres of a standard solution of lead acetate to 100 c.c. of distilled water in another Nessler glass and passing the gas through it. This method is satisfactory for all potable waters; but in the case of aerated waters containing vegetable acids, sugar, or organic matter, it is advisable to adopt the method described by Dr. F. L. Teed (*Analyst*, xvii. 142). Those substances which would cause the reduction of potassium bichromate do not interfere with the production of lead sulphide, and Dr. Teed's process also provides against the precipitation of copper and iron, both of which may be present in small quantities in aerated beverages. He proceeds as follows :

The liquid to be tested is placed in a Nessler glass, and a few c.c. of ammonia and a little potassium cyanide are added, and then a small quantity of ammonium sulphide. Any coloration then produced is due to lead alone, copper and iron not interfering. The lead sometimes found in aerated waters may be due to the metal-work in the case of siphons, but is apparently sometimes due to the action of the liquid on the glass. When such beverages (soda-water, ginger-beer, lemonade) are examined under the Food and Drugs Act, it is essential that three bottles should be purchased and opened, their contents mixed in a jug in the presence of the vendor, and then poured back into the bottles and sealed up.

In testing aerated, and, indeed, ordinary potable water, for lead, it is well to confirm its presence by the bichromate test, which must in the case of beverages be performed after the destruction of the organic matter present, by evaporating in a porcelain dish and igniting very gently with a drop or two of strong nitric acid to destroy carbon. The residue is then taken up with water, and a few drops of nitric acid and potassium bichromate added. A comparison can then be instituted, using a standard lead acetate solution treated in the same way.

Copper is not found in natural waters in this country, except possibly in the neighbourhood of copper lodes, but is common in some countries—for example, Barbadoes.

It is a disputed point how far copper is a dangerous poison in small quantities, and it is a more common constituent of many foods (such as wheat, cocoa, oysters, etc.) than is generally supposed.

It is probable that copper is less dangerous than lead, but no water should be passed as satisfactory which contains as much as $\frac{1}{30}$ grain of the metal per gallon.

The power of certain waters to attack lead is considerable; and while very soft waters are most notable for their plumbo-solvency, there are several causes which conduce to the action of water on lead.

Pure water, *e.g.* distilled water, has a decided action on lead, particularly if the lead is sometimes exposed to air, as might be the case in an intermittent service. This action is prevented if even small quantities of calcium salts and silica are present. Again, the presence of large quantities of nitrates or of chlorides is said to give water the power of acting on lead.

Probably the presence of a large quantity of carbonic acid and of vegetable acids, such as occur in a surface-water collected in moorland districts where peat abounds, is the most important cause of plumbo-solvency. In such waters it is likely to be found that the action is greatest in the autumn, when the decaying vegetation imparts a greater quantity of carbonic acid and vegetable acids to the water than at other times of the year.

Many widespread outbreaks of lead-poisoning have occurred in the past in districts supplied with moorland surface-waters; but the reported cases have very greatly diminished of late years, as the lead pipes were gradually replaced by tin, tin-lined lead pipes, or iron pipes.

A valuable and detailed examination into the causes of plumbo-solvency was carried out by Dr. Power, Dr. Houston, and others, at Settle, and will be found in the Report of the Medical Officer of the Local Government Board, 1893-4. In their investigations a very important point was clearly shown—namely, that the plumbo-solvency of a water collected from a peaty ground is influenced largely by the length of time the water stands in peat; if the water sinks into the ground and reappears again in the form of springs it loses its power of action on lead, whereas that water which comes directly from water-logged peat acts on lead in a marked manner.

An outbreak of lead-poisoning in Sheffield, which is supplied with a water from a moorland collecting ground, was successfully dealt with by artificially hardening the water by passing it over limestone; this plan is still in operation, and has since been adopted elsewhere.

In the investigation above referred to, Dr. Houston succeeded in isolating from peat infusion two bacteria, which he has denominated 'O' and 'Q,' which, when grown in peaty water, produce a considerable degree of acidity, and cause the water to act on lead. As regards the quantity of lead that should be permitted in drinking-water, the facts that lead is a cumulative poison, and that the plumbo-solvency of a water will probably vary with the season, point very strongly against the advisability of passing a water containing even the smallest traces. Phosphate of lead is very insoluble, so that if water containing lead is passed through an animal charcoal filter (which, of course, contains calcium phosphate) all lead will be removed from the water, until all the calcium phosphate is decomposed, when the filtering material must be renewed.

Determination of Organic Matter.—The determination of the organic matter—that is to say, the extent of the animal or vegetable pollution—is the most important factor in the sanitary analysis of water.

The three most important processes in use for the determination of the organic impurity in water are the following :

(a) The free and albuminoid ammonia process.

(b) The moist combustion or 'oxygen consumed' process.

(c) The combustion process of Frankland and Armstrong.

The Free and Albuminoid Ammonia Process.

This process was devised by Wanklyn and Chapman in the year 1867, and is now in use by water analysts all over the world. The object of this process is to obtain an indication of the amount of organic matter in a given water, by first estimating the amount of free or saline ammonia, and then determining the 'albuminoid' ammonia by boiling with a strongly alkaline solution of potassium permanganate. This decomposes the organic matter present in the water with the evolution of a further quantity of free ammonia.

Estimation of Free or Saline Ammonia.—The process is usually worked as follows: 500 c.c. of the water to be

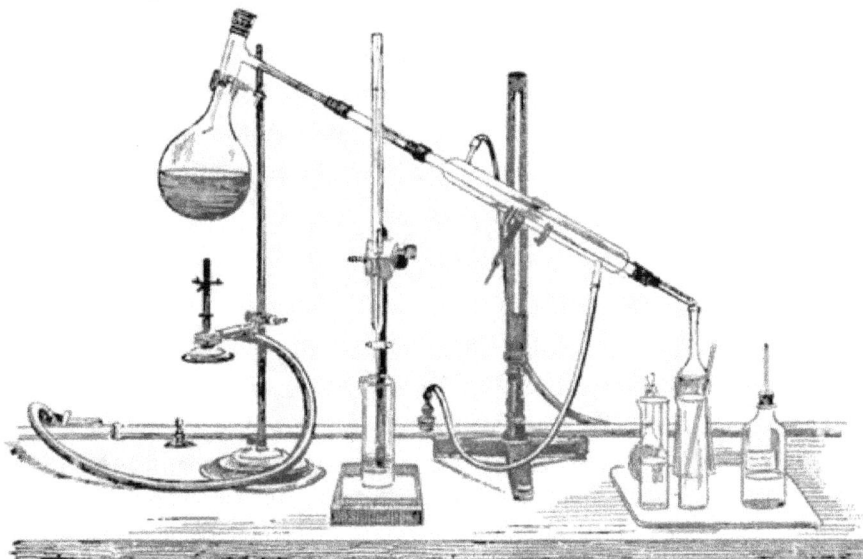

APPARATUS USED IN WANKLYN'S PROCESS.

examined are placed in a retort or flask capable of holding at least 1200 c.c., and connected to a Liebig's condenser. A side tube flask is preferable to a retort, as it is far more difficult for splashings to pass into the distillate than in the case of a retort. A small quantity of sodium carbonate (about $\frac{1}{2}$ grain) is added to the water to ensure the liberation of any ammonia that may be present as chloride, sulphate, nitrate, or any other non-volatile salt. As ammonia salts are the commonest impurity in carbonate of sodium, care must be taken to use only sodium carbonate that has been previously ignited. Sometimes this addition is not necessary, owing to the water being slightly alkaline naturally, but it is preferable not to omit it in any case. It is customary to distil over one-tenth, that is, 50 c.c., into a Nessler glass which has previously been rinsed with ammonia-free water. As a matter of experience, it is found that in the case of most waters three-quarters of the total saline ammonia will be present in the first tenth, and it is therefore customary to add on one-third of the ammonia found in the first cylinder, and to take this as representing the whole saline ammonia rather than to nesslerize the succeeding distillates.

We prefer, instead of this, to collect the first 100 c.c. in a 100 c.c. flask, and, after careful mixing, to nesslerize one-half of this, keeping the other portion in reserve. The quantity of ammonia so found is, of course, one-half of the total saline ammonia. The reason for preferring this course is that, in the case of some waters, so much ammonia is yielded by the first 50 c.c. of distillate that it is impossible to estimate it by nesslerizing, so that by collecting 100 c.c., and nesslerizing half of this, we have less ammonia to deal with; and if this still proves too much, as is sometimes the case, we can take a measured portion of the remaining 50 c.c., dilute it up to 50 c.c. with ammonia-free water, and make an estimation on this.

In the case where 50 c.c. are distilled off first, it is customary to distil off and throw away three other lots of 50 c.c., thus leaving 300 c.c. in the retort; so that if we distil off 100 c.c. at first, as above suggested, then a further 100 c.c. should be distilled off and discarded, leaving 300 c.c. in the retort, as in the ordinary manner of carrying out the process.

The operation of determining the amount of ammonia in

the distillates, or 'nesslerizing,' as it is technically called, is carried out as follows:

To 50 c.c. of the liquid to be tested, which is placed in a Nessler glass, add 2 c.c. of sensitive Nessler reagent, and mix the liquid thoroughly by a rotary motion (the pipette is not to be used as a stirring rod, as this will serve to dilute the Nessler reagent next time the pipette is used for its legitimate purpose). If the liquid contains the least trace of ammonia, a yellowish-brown colour will be produced, which within certain limits corresponds precisely to the amount of ammonia present. The reaction that takes place is: $NH_3 + 2HgI_2 + 3KHO = NHg_2I + 3KI + 3H_2O$. If more than ·0005 of a gramme of ammonia is present in the 50 c.c. of water that has been nesslerized, instead of a clear, coloured solution being produced, an actual precipitate will be formed, and it is then impossible to compare the colour against the standard solution as is desired. The colours are best compared by looking down through the tubes held an inch above a white tile or piece of paper. It is therefore essential that the Nessler glasses used for comparing two liquids shall be of precisely the same bore.

The liquids to be nesslerized must all be the same temperature, preferably about 15·5° C.

Standard Ammonia Solution.—The standard solution of ammonia generally employed is made up so that 1 c.c. shall contain ·00001 gramme of ammonia; the salt generally used is ammonium chloride (NH_4Cl). If we weigh out 3·15 grammes of the pure salt, and dissolve this in a litre of water, this will form a convenient stock solution from which to prepare an ammonia standard containing ·00001 gramme of ammonia per c.c., which we do by taking 10 c.c. of the strong solution and diluting it to 1000 c.c. with distilled water. It is hardly needful to say that the water used to make up these solutions must be ammonia-free. The following calculation shows why 3·15 grammes of ammonia salt are employed: $-NH_4Cl = 53·5$, $NH_3 = 17$, $\therefore$ 1 gramme NH_3 is contained in 3·15 grammes NH_4Cl.

Preparation of Nessler's Reagent.—Nessler's solution consists of a solution of mercuric iodide and potassium iodide, and is prepared by dissolving 35 grammes of potassium

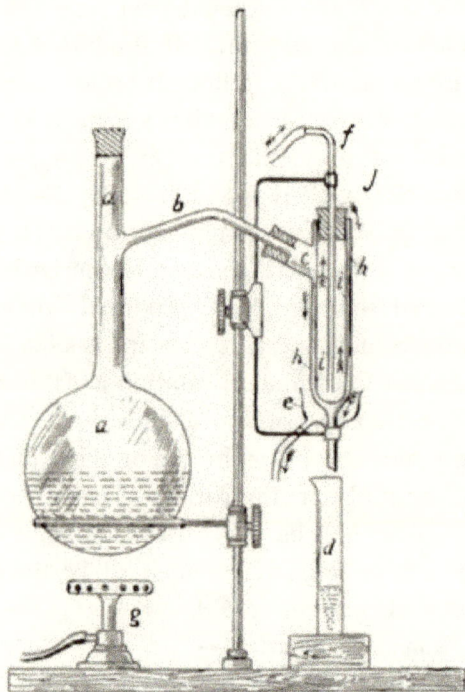

APPARATUS FOR WATER DISTILLATION.

iodide in about 100 c.c. of distilled water, and then adding to it a cold solution of 16 grammes of mercuric chloride dissolved in about 300 c.c. of distilled water, when it will be found that a slight permanent precipitate is produced. The mixture is then made up to 1000 c.c. by the addition of 20 per cent. caustic potash solution. Even after the solution has been allowed to stand at rest for some weeks, and the clear liquid drawn off, it will still continue to deposit a yellowish sediment; but this need cause no trouble if it is not stirred up by inserting the pipette right to the bottom of the bottle. The danger of sucking up the liquid, which is exceedingly unpleasant to taste, may be avoided by fitting into a rubber cork a pipette with its bulb at the end. As the bulb is constantly immersed in the solution, the required amount can be retained by simply placing the finger on the free end. If a long narrow bottle is used, the sediment will remain undisturbed at the bottom, especially if some fragments of glass are kept in the bottom of the bottle.

Nessler's solution is more sensitive after it has been kept for some time than when freshly made, and therefore it is customary to keep enough solution in stock to ensure its having been made at least a month before it is used. Even when the solution has been kept for a month or more before use some five minutes must be allowed to elapse before the full colour is obtained, and if fresh Nessler is employed the colour may take even longer to develop.

The above apparatus consists of a flask a, heated by a ring-burner g, a condenser h, i, and a Nessler glass d, with a 50 c.c. mark, to act as a receiver. The flask is one of the ordinary Würtz pattern, fitted with an indiarubber stopper, and connected by the cross-tube b to a condenser of the form recently described by Cecil H. Cribb in the ' Analyst' (May, 1898), which, while only occupying about one-third of the space, has a much greater efficiency than that of a Liebig's condenser. The steam enters at c, and passes into the very narrow annular space between the two tubes h and i, where it is condensed and runs down into the Nessler glass below. The condensing space is cooled on both sides by cold water flowing in the direction of the arrows, down the tube f, up through i, which it fills, and then passes out through slits in the cork j, to flow down the outside of h into the cup e, whence it runs away to waste.

Determination of the Albuminoid Ammonia.—The residue in the flask, after distilling off the free or saline ammonia, is allowed to cool somewhat, and 40 c.c. of alkaline permanganate are added and the distillation continued. In cases where a water contains a large amount of organic matter, as in peaty or highly-polluted waters, it may be necessary to add more alkaline permanganate, while in the case of sewage or sewage effluents a special procedure must be adopted. Under the action of alkaline permanganate the nitrogenous matter in the water, with the exception of that present in an oxidized form (nitrites or nitrates), is decomposed, the nitrogen being yielded in the form of ammonia. This decomposition does not occur instantaneously, and hence it is necessary to distil and collect four quantities of 50 c.c., which are each nesslerized separately, and the amounts of ammonia indicated added together and reported as 'albuminoid ammonia.'

Excessive 'bumping' in the distilling-flask may be best moderated by placing some pieces of pipe-stem in the flask.

Alkaline permanganate is prepared as follows : Dissolve 200 grammes of good caustic potash in a litre of ammonia-free distilled water previously heated to boiling; when it is dissolved add 8 grammes of potassium permanganate, and evaporate the liquid down to about one-quarter of its bulk. Now make up to about a litre again with ammonia-free water and evaporate as before. These two evaporations will expel any ammonia that may exist in the liquid, or that may be formed from traces of cyanide in the caustic potash. After the second evaporation, make up to a litre with ammonia-free water and keep in a stoppered bottle.

Ammonia-free water is best prepared as follows : A little sulphuric acid is added to ordinary distilled water in a retort, and the distillate is collected after rejecting the first 100 c.c. The next 600 or 700 c.c. will be free from ammonia, and the distillation should then be stopped, or traces of ammonia will again begin to come over. It is not, however, worth while to go to the trouble of preparing it one's self, as distilled water can now be purchased free from ammonia at 3d. a gallon, which is cheaper than it could be made in the laboratory.

Before applying Wanklyn's process to a water, it is advisable to nesslerize 50 c.c. of the sample, and if any perceptible colour beyond a very faint yellow is produced, a less quantity than 500 c.c. should be taken and made up to 500 c.c. with ammonia-free water. In the case of a river-water, for example, 100 c.c. would be a convenient quantity, while in the case of a sewage or sewage effluent 10 or 20 c.c. will be found to contain as much ammonia as can conveniently be nesslerized. When working on such small quantities any error that may occur in manipulation is obviously magnified by calculation, and hence *it is preferable not to employ the process of nessleriza-tion at all in the case of sewage and sewage effluents, but to work on the full quantity of 500 c.c., and to titrate the ammonia in the distillate with decinormal acid, using cochi-neal as the indicator.*

If the 'free' ammonia does not exceed 0·003, and the 'albuminoid' 0·007, part per 100,000 respectively, the water may be generally regarded as within the limit of safety. A high ratio of 'albuminoid' compared with the 'free' denotes vegetable pollution, in which case the amount of chlorides and nitrates present will be low. Much 'free' ammonia accompanied with excess of chlorides will indicate the presence of animal contamination.

Wanklyn's standard for albuminoid ammonia is as follows :

High organic purity	...	·0 to ·05 part per 1,000,000
Satisfactory purity	...	·05 to ·10 part per 1,000,000
Polluted	...	Over ·10 part per 1 000,000

Wanklyn states that when the albuminoid ammonia amounts to 0·05 part per million, then the proportion of free ammonia becomes an element in the calculation, but if it is absent, then the water may be passed as organically pure despite much free ammonia and chlorides. If free ammonia is present in but very small amount, a water should not be condemned unless the albuminoid ammonia is about 0·10 part per million.

A certain amount of information may be gained by noting the rate at which the ammonia is evolved during the distillation. Rapid evolution denotes that the organic matter is in a decomposing or putrescent condition, whereas if the ammonia is given off gradually it indicates that the contained organic matter is derived from comparatively recent pollution.

It is worthy of note that if the original sample on being tested with Nessler's reagent gives the faintest colour, it shows the presence of an objectionable amount of free ammonia, and the sample is open to grave suspicion.

It is important to note that deep well waters, particularly those in the chalk and greensand, frequently contain a large proportion of free ammonia, but such waters will be practically free from any trace of albuminoid ammonia. A high amount of free ammonia is of course also found in rain-water, and frequently also in water that has percolated through soil containing reducing salts of iron, which decompose the nitrates originally present in the water. This change also takes place in water that has stood in iron pipes.

The Moist Combustion or 'Oxygen Consumed' Process.

This process, which was originally due to Forchammer, has been modified by Tidy, Dupré, and others.

The object of the process is to ascertain the amount of

organic matter in solution in a given water by finding how much oxygen is absorbed in a given time at a given temperature from an acid solution of potassium permanganate. It does not distinguish between organic matter due to sewage and that due to vegetable débris; nor is the reaction so immediate or complete as to make it possible to titrate a given quantity of sample water direct with permanganate, as was attempted when the process was first introduced.

The amount of permanganate used up by a water will vary with the temperature at which the mixture is kept, the length of time it is allowed to act, and, above all, with the precise substances which the water contains.

It will therefore be manifest that the process is an arbitrary and not a precisely quantitative one, and in order to obtain concordant results we must observe the same procedure as to time, temperature, and reagents in every case.

The process depends upon the fact that potassium permanganate, in the presence of H_2SO_4, gives up five-eighths of its oxygen to any oxidizable matter which may be present. The reaction which takes place is :

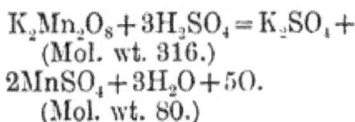

$$K_2Mn_2O_8 + 3H_2SO_4 = K_2SO_4 +$$
(Mol. wt. 316.)
$$2MnSO_4 + 3H_2O + 5O.$$
(Mol. wt. 80.)

It will be seen that 316 parts by weight of potassium permanganate yield 80 parts of oxygen, or one part of oxygen will be yielded by $\frac{316}{80}$ ($= 3\cdot95$) parts of potassium permanganate. The reagents required will be :

(1) Potassium permanganate of such a strength that 1 c.c. will contain ·1 milligramme of available oxygen. This is made by dissolving ·395 gramme of potassium permanganate in a litre of organically pure water. (As we

5

want ·1 milligramme of O in a cubic centimetre, we shall require ·1 gramme in a litre; then, if 316 grammes of potassium permanganate contain 80 grammes of available oxygen, ·395 gramme of potassium permanganate must be dissolved in a litre of water.)

(2) Sodium hyposulphite solution, 1 gramme of $Na_2S_2O_3$ to 1 litre.

(3) Starch solution, about 2 per cent.

(4) Potassium iodide solution, about 10 per cent.

(5) Dilute sulphuric acid 1 : 3, in water free from organic matter.

The method most commonly preferred by analysts in this country, and the process recommended by the Society of Public Analysts, is to determine the amount of oxygen consumed at a temperature of 80° F. in two separate experiments, viz., in fifteen minutes and in four hours.

The amount of oxygen consumed in fifteen minutes will represent very easily reducible organic matter, and will include any nitrites, sulphides, and ferrous salts present in the water under examination. The four-hours figure will represent the whole of the organic matter contained in most waters, but with very bad waters even more time is required for the complete oxidation of the organic matter. When this process is applied to water containing much suspended matter, it is certain that a considerable proportion of organic matter, such as fibres of paper, remain unacted on.

Determination of the 'Oxygen Consumed' in Fifteen Minutes.—Into a flask of 500 c.c. capacity place 250 c.c. of the water to be examined. Into another flask place 250 c.c. of distilled water, and heat on the water-bath to 80° F.; when this temperature is recorded, pour into each flask 10 c.c. of the potassium permanganate solution and 10 c.c. of dilute sulphuric acid. Maintain the solutions at 80° F. for fifteen

minutes. If the pink colour becomes faint, a further quantity
of permanganate solution must be added to each flask. At
the end of the fifteen minutes the amount of potassium permanganate, in both the standard (or 'blank') and the sample,
has to be estimated. The sample will not be so bright a
colour as the standard, because some of the permanganate
is used up by the organic matter. The next step is to
estimate the residual permanganate. This is conveniently
effected by adding to each flask some solution of potassium
iodide, when the remaining permanganate sets free from
the potassium iodide an amount of iodine exactly corresponding to the amount of permanganate unused; thus:

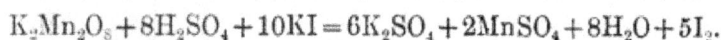

$$K_2Mn_2O_8 + 8H_2SO_4 + 10KI = 6K_2SO_4 + 2MnSO_4 + 8H_2O + 5I_2.$$

On the addition of the potassium iodide the pink colour
of the solution will change to a yellow, due to free iodine,
which dissolves in the excess of potassium iodide, and this
can be conveniently estimated by titration with sodium
hyposulphite. A solution of iodine in water at a temperature of 80° F. is very sensibly volatile, so that the liquid
should be well cooled before adding the potassium iodide;
if this is not done, purple vapours and the characteristic
smell of iodine will be noticed, and an inaccurate result will
be obtained. This solution reacts with the liberated iodine
according to the equation:

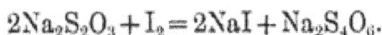

$$2Na_2S_2O_3 + I_2 = 2NaI + Na_2S_4O_6.$$

The hyposulphite solution is run into the iodine-tinted
liquids from a burette until the tint has entirely disappeared. The end of the reaction is best perceived by
adding a few drops of starch solution to the liquids; this
forms a blue compound with the iodine. As soon as the
iodine has been used up by further titration, the blue colour
will disappear; and we have now all the data necessary for
estimating the amount of oxygen absorbed.

Example of Calculation—(*a*) *Sample.*—Maintained at 80° F. for fifteen minutes, 10 c.c. of potassium permanganate added, and then KI, and titrated with hyposulphite, of which 12 c.c. were required.

(*b*) *Distilled Water.*—Treated as sample, and found to require 20 c.c. of hyposulphite solution.

Therefore, 20 c.c. of the hyposulphite solution are equivalent to 10 c.c. of the permanganate solution, or to 1 milligramme of available oxygen. Therefore, 1 c.c. of $Na_2S_2O_3 = .05$ milligramme of oxygen.

The sample required only 12 c.c. of the $Na_2S_2O_3$ solution, while the distilled water required 20 c.c., so the difference is 8 c.c.

Therefore, an amount of oxygen has been absorbed equivalent to 8 c.c. of the $Na_2S_2O_3$ solution; but it has been seen that 1 c.c. of the $Na_2S_2O_3$ solution is equal to .05 milligramme of oxygen. Therefore, 8 c.c. is equivalent to .40 milligramme of oxygen.

Therefore, .40 milligramme of oxygen is absorbed by 250 c.c. of the sample water, or 1·6 milligramme is absorbed by 1000 c.c of oxygen.

Therefore, the oxygen-consuming power of our water is 1·6 milligramme per litre, or 1·6 part per million, or 0·16 part per 100,000.

Determination of Oxygen Consumed in Four Hours.—This is carried out as above, the flasks being allowed to remain at a constant temperature of 80° F. for four hours in a specially constructed water-bath or incubator. If the pink colour of the solution should become faint, a further 10 c.c. of the permanganate solution should be added. When the proper time has elapsed, the solutions are titrated with the hyposulphite solution, as directed above.

McKay's Modification of the Oxygen Consumed Process.—L. W. McKay (*Chemical News*, xlvii. 195) titrates the

permanganate with ammonio-ferrous sulphate. In using
Tidy's permanganate method for estimating the organic
purity of waters, that author has always experienced a
difficulty in ascertaining the precise moment of the dis-
appearance of the blue colour of the iodized starch; to
avoid this he has adopted the use of ammonium ferrous
sulphate. The solutions he employs are: (1) 0·395 gramme
permanganate in 1000 c.c. pure water; (2) 4·90 grammes
ammonium ferrous sulphate in 975 c.c. water and 25 c.c.
concentrated sulphuric acid. From several experiments,
he feels sure that the method is very good; the results
are not only constant, but also agree well with dupli-
cate analyses done by Tidy's method. The advantages
claimed for the method are: (1) The abolition of the blue
colour difficulty; (2) saving of time; (3) two solutions only
are required; (4) the amount of chemically pure water
necessary is reduced to a minimum. The ammonium
ferrous sulphate solution keeps very well in the dark. We
have used this method, and find it gives a sharper end-
reaction than the ordinary method.

The estimation of the oxidizable organic matter by means
of acid permanganate gives exceedingly valuable data when
taken in conjunction with the foregoing process of Wanklyn.
Taken by itself, the determination of the oxygen required
to oxidize the organic matter is liable to furnish misleading
results, owing to the fact that some forms of organic
matter reduce less permanganate than others. Thus, a
water derived from a peaty upland is liable to absorb
a larger amount of oxygen than one containing a very
serious amount of dangerous pollution. This, which at
first sight appears to detract greatly from the value of the
process, is a matter of but little consequence if the source
of the water is known. It should not be overlooked that
the presence of lower oxides of iron, nitrites, and sulphur

compounds other than sulphates also absorb oxygen ; and this should be allowed for if these bodies are present. The following table of approximate standards for this process has been drawn up by Frankland and Tidy :

AMOUNTS OF OXYGEN ABSORBED BY 100,000 PARTS OF WATER.

	Derived from upland surfaces.	Derived from sources other than upland surfaces.
Water of great organic purity	Not more than 0·1	Not more than 0·05
Water of medium purity	,, ,, 0·3	,, ,, 0·15
Water of doubtful purity	,, ,, 0·4	,, ,, 0·2
Polluted water ...	More than 0·4	More than 0·2

Hot Moist Combustions.—Abroad it is the practice to perform the permanganate determination hot; and some authors find advantage in doing it in alkaline instead of acid solution. So far as the temperature is concerned, there is no doubt that a larger proportion of organic substances are attacked, and this in a considerably shorter time, by permanganate at boiling-point than at 80° F. The principal objection to its use, and one which hitherto has almost prevented its adoption in this country, is the liability of chlorides to interfere with its results. Experiments of Preusse and Tiemann have shown that up to 0·4 per cent. of sodium chloride—that is to say, for all ordinary amounts of chlorides —this objection does not necessarily arise. At the same time these authors admit that in the acid process it is liable to occur presumably through the decomposition by chlorine of the organic constituents which are present in some waters. They state, however, that it does not occur with the alkaline process, and accordingly recommend the latter for those cases where the chlorides appear to produce a disturbing effect. There is this advantage in performing both estimations, that most organic substances are affected to a different extent by the alkaline and by the acid permanganate process ; and the twofold determination may give, therefore, two factors which may in some measure be independent of each other, and of which the separate estimation might possibly be valuable in systematic periodical examinations, which it must in season and out of season be remembered constitute the really valuable contribution of chemistry to sanitary science. Thus Pouchet and Bonjean found in 1,127 waters examined chemically and bacteriologically in the laboratory of the Consultative Committee of Public Health of France, that in those samples (297 in number) which absorbed more than 1 milligramme of oxygen, a larger amount of oxygen was absorbed in alkaline than in acid solution; and that of those 297 waters, 251 were found both by direct inquiry and bacteriological observation to be grossly contaminated by fæcal matters. A large number of special experiments on solutions artificially prepared, etc., show that no invariable rule can be laid down to the effect that animal matters are more attacked in alkaline than in acid solution. In some cases, but not in all, the difference may have been attributable to the presence of chlorides, which, as stated by Tiemann, are not liable to interfere with the reaction in alkaline solution. Whatever the cause, it appears clear that the relation in question occurred in the majority of instances, and probable that the double determination will frequently be instructive. It is the less to be avoided because, on whatever process it is conducted, the apparatus is simple, and the personal time involved is not large.

The hot process most used abroad is that of Kubel, who boils for five or ten minutes in acid solution. Others, as Drown and Hazen, bring the solution to the boil before adding the water. Schulze performs the reaction in permanganate with caustic soda. On the whole, the process most to be recommended is that adopted in the laboratory of the Consultative Committee, as described in the paper Pouchet and Bonjean to which we have referred.

Of four conical flasks, two receive 100 c.c. each of the water under examination and two receive 50 c.c. each. One 100 c.c. is acidified with 10 c.c. of sulphuric acid 1 : 4, and one 50 c.c. with 5 c.c. of the same solution. The other 100 c.c. receive 10 c.c. of a saturated solution of sodium bicarbonate (the action of caustic soda or potash on the permanganate being considered too strong), and the other 50 c.c. receive 5 c.c. of the same solution. Ten c.c. of a 0·5 per 1000 solution of permanganate is then added to each flask, and the whole are brought to the boil and kept so for ten minutes. The flasks are then allowed to cool, and the alkaline solutions acidified with 20 c.c. and 10 c.c. respectively of sulphuric acid (1 : 2) solution. Ten c.c. of ammonium ferrous sulphate solution, containing 10 grammes of the salt and 10 grammes of sulphuric acid per litre, are then introduced, and the solution is titrated with the permanganate solution until a faint pink tint is produced. The difference between the results of the 100 c.c. and the 50 c.c. in each case gives the figure for the permanganate required for the oxidation of the organic matter in acid and in alkaline solution respectively.

The permanganate solution is titrated with a solution of pure crystallized oxalic acid, the purity of which is gravimetrically verified by transformation into calcium oxalate and then carbonate. Special care is taken to avoid dust; use scrupulously clean flasks, and rinse with pure and freshly-prepared distilled water. The results are expressed either in terms of oxygen absorbed by 1 c.c. of permanganate, corresponding to 0·1266 milligramme of oxygen, or of organic matter as oxalic acid, 1 c.c. of permanganate corresponding to 0·997 milligramme of crystallized oxalic acid :

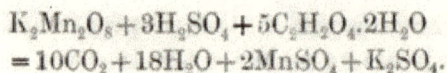

$$K_2Mn_2O_8 + 3H_2SO_4 + 5C_2H_2O_4.2H_2O$$
$$= 10CO_2 + 18H_2O + 2MnSO_4 + K_2SO_4.$$

It must be borne in mind that equal weights of different organic materials absorb entirely different quantities of oxygen from permanganate solution under any treatment; the results of this and any similar method, even for water from the same source, determine quite arbitrary and undefined constituents. That these determinations nevertheless are of value has been shown repeatedly by the comparison of results which they have yielded with those of direct examination of the source. It is clear, however, that, more even than the choice of process, the rigorous and minute uniformity of procedure in all cases is necessary for getting the best use out of the determinations.

A still higher proportion of organic matter can be oxidized by the potassium bichromate process of Wolff-Degener-Herzfeld, but at the cost of more complicated apparatus and a much longer process.

It is to be remarked in regard to all these processes, as well as to the actual application of ultimate analysis in the ingenious method of E. Frankland, that the nearer approximation to completeness in the amount of organic carbon and nitrogen has a value which is more apparent than real. The significance of these elements depends on the manner in which they were combined in the substances which were dissolved in the water; and in the absence of this information it is more than doubtful whether the search for exact organic carbon and nitrogen figures is worth great pains· For our own part, when account is taken of these circumstances, we feel persuaded that more useful information is to be obtained in regard to a water by moist combustions and albuminoid ammonia determinations than by a single ultimate organic carbon and organic nitrogen determination, which would occupy about the same time and be subject to corrections of such extent as to make the accuracy of the final figures very liable to be erroneous.

On this ground we do not give a detailed description of the Frankland process, which can be found either in that author's 'Water Analysis' or in Sutton's 'Volumetric Analysis.' As, however, it is the official process for some purposes, we subjoin a description which will enable the results obtained by it to be critically followed.

The Frankland Combustion Process.

As soon as possible after the receipt of the water, and directly after the estimation of the ammonia, a quantity of water, varying from 1 litre (where the ammonia is less than 0·05 part per 100,000) to 100 c.c. (where it is more than 1·0 part), is boiled briskly with saturated sulphurous acid solution, so decomposing carbonates, nitrates and nitrites, expelling CO_2, and fixing the larger part of the ammonia, except for a small loss, to correct for which empirical tables are provided. In the final calculation the nitrogen of the total ammonia found in the previous determination, less the empirically estimated loss, is deducted from the total nitrogen yielded by combustion. The acidulated water is evaporated over a water-bath, a drop of ferrous chloride solution being added to destroy nitrates, and a little sodium sulphite, if necessary, to fix free sulphuric acid. The evaporation, which is conducted under a glass shade to rigidly exclude dust, takes about eighteen to twenty-four hours for a litre; and if nitrates and nitrites are high, the solid residue has then to be treated with more sulphurous acid solution, and again evaporated to dryness. With liquids containing much ammonia and no nitrates or nitrites, ammonic phosphate (which is inapplicable in the presence of the latter salts, because it does not reduce them) is recommended, instead of sulphurous acid, as the loss of ammonia is stated to be less. The dry residue is mixed with copper oxide, for the greater part finely divided, placed in a narrow combustion-tube between coarse copper oxide, followed on the further side by copper gauge-cylinders or turnings, and then by more copper oxide, exhausted with a mercury-pump, and combusted in the ordinary way. The resulting CO_2, N_2O_2, N, and occasionally CO, are drawn into a special eudiometer, and determined in the usual way. In addition to the correction for loss of ammonia, which varies within wide limits with the total quantity in the water, it is found that corrections must be made for carbon and nitrogen introduced accidentally during the analysis; and for this purpose several blanks, each on the evaporation of 1 litre of water, free not only from ammonia, but also from organic matter (a substance extremely difficult to obtain, and still more so to keep), are recommended as desirable. It is to be observed also that this process takes no account of the organic substances occurring in natural waters which are volatile in acid solution.

The authors of this process believe that, with certain reservations, the proportion of organic carbon to nitrogen indicates whether the polluting organic matter was derived from animal or vegetable sources. Much carbon and little nitrogen indicates vegetable contamination, whereas, if the ratio of carbon to nitrogen is low, it is evidence of animal pollution.

The Rivers Pollution Commissioners found that in peaty waters the ratio of nitrogen to carbon was 1 : 11·9, while in similar waters that had been stored in lakes the nitrogen to carbon = 1 : 5·9. In sewage, the average of a large number of samples gave nitrogen to carbon = 1 : 2·1. Highly polluted well-waters gave nitrogen to carbon = 1 : 1·3.

The Rivers Pollution Commissioners state that 'a good water should not yield more than 0·2 parts of "organic carbon," or more than 0·02 of "organic nitrogen" in 100,000 parts.'

Oxidized Nitrogen.

Nitrites and Nitrates.—When nitrogenous organic matter undergoes decomposition ammonia is invariably produced, and by oxidation this ammonia is converted by the agency of micro-organisms, which exist in most natural waters and soil, into water and nitric acid, giving rise to nitrates of sodium, potassium, and calcium.

When a water suffers pollution with animal or vegetable matter, the tendency under natural conditions is towards purification by oxidation. This oxidation of the organic matter may take place in several ways, the most important of which are the following : (a) Eremacausis, (b) fermentation, and (c) putrefaction.

The process of eremacausis is a slow breaking or burning up of the organic matter by means of slow combustion, which may be likened to the process of phosphorescence. The process of putrefaction is due to the vital activity of certain bacteria, whereby the organic matter is first resolved into simpler compounds, and eventually into their simplest elements. Probably the most important process concerned in the breaking up of organic matter is the process of fermentation known as nitrification, due to certain organisms which oxidize ammonia into nitrites, and to others which convert the nitrites into nitrates.

It was first pointed out by Frankland that nitrates when present in quantities sensibly above what have been found to be 'normal' for the particular source in question, must be regarded as due to 'past pollution.' The presence of small quantities does not necessarily denote the presence of organic pollution, since nitrates always exist in waters of the highest organic purity derived from the chalk and lias.

The estimation of nitrates and nitrites in a sample of water, therefore, gives us the measure of the oxidized organic matter which previously existed in the water, or, in other words, the past pollution.

When organic nitrogen is fully oxidized to nitric acid it may be held to be quite innocent. Still, it indicates past pollution, and if its quantity exceeds a certain amount, it calls for a very close investigation of the antecedents of supply in question.

Thus it will be seen that small quantities of nitrates derived from deep wells in these formations possess no particular significance, as they are probably derived from organic matter that has long since become oxidized, and which was originally derived from the surface, or possibly from the fossil remains.

Waters containing more than 0·2 part per 100,000 are open to suspicion unless they can be accounted for as being derived from the strata. Where this source of nitrates cannot be excluded, a greater quantity than 0·5 part per 100,000 should be regarded with very grave suspicion, as the quantity of nitrates derived from unpolluted sources very rarely exceeds 0·3 part per 100,000; any larger quantity than this is almost certain evidence of pollution.

Nitrites.—The presence of nitrites is generally to be regarded as a bad sign. When organic matter is oxidized into nitrates, nitrites are first formed, and these are afterwards oxidized into nitrates. Thus it will be seen that the presence of nitrites indicates the probability that the organic pollution is very recent, owing to the fact of the oxidation being incomplete.

It may be as well to point out that the particular combination in which both the unoxidized and oxidized nitrogen exist in samples of water are very liable to undergo change from the moment of collection. Hence it is not to be wondered at that different analyses of the same sample may show very considerable differences in the figures obtained, unless the samples have been examined almost immediately after collection.

It is very possible that the formation of nitrites is intermediate between the oxidation of ammonia and the formation of nitric acid, while nitrites are so rarely found on account of their great instability and the ease with which they are further oxidized to nitrates.

Various observers appear to have met with nitrites in waters intended for domestic purposes, but their occurrence is far more uncommon than is usually believed, and in many recorded cases may have been due to impure reagents, or to water having been kept some time before analysing. They would probably only be found in waters in which nitrates were in company with large quantities of organic matter which was abstracting oxygen from them. Nitrites may be shown to exist in large quantities by the reduction of nitrates in the copper-couple process if the liquid is tested before the reduction into ammonia is complete.

Nitrites are formed in well-water rich in nitrates, and in sewage-effluents containing nitrates, if the water or effluent is kept in closed bottles. Whether this is due to the action of bacteria or merely to a purely chemical reduction might possibly be decided experimentally.

Determination of Nitrates.

Many different methods have been proposed for the estimation of nitrates, but only three are at all generally employed.

The various methods that have been used or suggested may be divided into three classes :

(1) Methods in which the nitrates are reduced to ammonia and estimated as such by nesslerizing, or, if present in great quantity, by titration.

(2) Colorimetric methods.

(3) A method in which the nitrogen of the nitrate is evolved as nitric oxide gas, which is measured.

Of the first group the copper couple method is worthy of note as being very easily performed and giving accurate results.

There are several colorimetric methods, such as the indigo test, the brucine test, and the phenyl-sulphate method. The latter is commonly employed on account of its yielding fairly accurate results in a short space of time.

Qualitative tests for the presence of nitrates in water are obviously of no use, seeing that nitrates are found in varying quantities in almost every water except rain-water.

The only method representative of the third group, in which the nitrates are converted into nitrogen gas, is that of Crum, which is now becoming superseded by the phenyl-sulphate method.

Estimation of Nitrogen as Nitrates and Nitrites by Reduction to Ammonia.

Copper-Couple Method.—When a piece of zinc coated with copper is placed in water it slowly decomposes it, hydrogen being set free and the oxygen uniting with the zinc to form zinc oxide. If nitrogen be present as nitrates (or nitrites), the hydrogen at the moment of its liberation first reduces the nitrates to nitrites, and then unites with the nitrogen to form ammonia, and the ammonia so formed can be estimated by means of Nessler's test. On this is based a method for the estimation of nitrates in water. For the estimation we require a bottle holding about 200 c.c., a piece of zinc-foil, and a 3 per cent. solution of copper sulphate. Take a piece of zinc-foil about 4 in. long and 2 in. broad, clean it with HCl, then immerse it in a 3 per cent. solution of copper sulphate. The copper will be deposited as a black coating on the zinc. It must not be kept in the solution too long; about ten minutes is enough. The use of granulated zinc is advocated by W. F. K. Stock (page 107, *J.S.C.I.*, 1897), as offering a greater surface; and after the estimation the zinc is kept in the bottle for the next occasion, and is again available after re-coppering.

Transfer the copper couple to a bottle of about 200 c.c. capacity and wash it well with distilled water, then rinse out with the water to be examined, and when washed fill up the bottle with the water to be examined, and leave in the dark for twenty-four hours. The nitrates in the water will be converted into ammonia and remain in solution as such. The next morning the amount of ammonia formed may be determined. Williams (*Analyst*, p. 1881, 36), to whom this ingenious and exact process is due, states that the reaction may be hastened by slightly warming the solution. He also states that the rapidity of the reaction may be further increased by the addition of about 0·1 per cent. of salt. Williams has shown that nitrous acid always remains in the solution until the reaction is finished. By testing for nitrous acid the completeness of the reaction can be obtained with certainty. The most delicate test which can be applied for this purpose is Greiss's metaphenylenediamine test (*v. infra*). After shaking, take out from 5 to 10 c.c., according to the purity of the water, dilute with about 100 c.c. of ammonia-free water, distil off 50 c.c. into a Nessler glass and nesslerize in the usual manner. The results are calculated as 'nitrogen,' after previously subtracting the amount of ammonia which was found to be present as free ammonia. This correction need only be applied to waters containing much saline ammonia.

Almost as good results can be obtained after a little experience without distillation by nesslerizing the water direct after treatment by the zinc-couple method. Five to ten c.c. of the water is diluted to 50 c.c. with ammonia-free water and nesslerized as usual. A little experience is necessary to make proper allowance for the turbidity produced by the calcium salts acting on the Nessler reagent.

Blunt (*Analyst*, vi. 202) points out that the above process may be used without distillation and with accuracy in the case of any water by adding a small quantity of oxalic acid to a double quantity of the water, dividing, and using one portion, clarified by subsidence, as a comparison liquid for testing against the other, which has been acted upon by the zinc-couple couple.

Where dilution is necessary, both portions must be diluted equally. This plan possesses the advantage that an equal turbidity is produced by Nessler reagent in both portions.

Phenyl-sulphate Method for the Estimation of Nitrates in Potable Water.

The constitution of phenol-sulphonic acid is shown by the equation :

$$C_6H_5OH + H_2SO_4 = C_6H_4(OH)SO_3H + H_2O.$$

The reaction that occurs when this reagent (which, when made up as described below, contains phenol-sulphonic acid and an excess of sulphuric acid) is added to a nitrate is as follows :

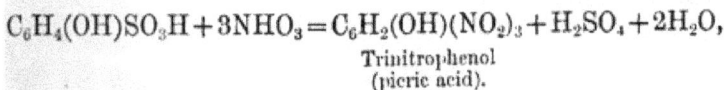

$$C_6H_4(OH)SO_3H + 3NHO_3 = C_6H_2(OH)(NO_2)_3 + H_2SO_4 + 2H_2O,$$

Trinitrophenol
(picric acid).

and on addition of ammonium hydrate :

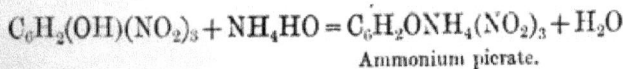

$$C_6H_2(OH)(NO_2)_3 + NH_4HO = C_6H_2ONH_4(NO_2)_3 + H_2O$$

Ammonium picrate.

This very quick and convenient process is carried out as follows :

6

Evaporate 50 c.c. of the sample to dryness in a platinum dish on the water-bath, then add 1 c.c. of phenylsulphate (made by mixing together 18·5 c.c. of strong sulphuric acid, 1·5 c.c. of water, and 3 grammes of phenol), mix with a rod, and allow to remain on the water-bath for three minutes. Add about 1 c.c. of water, about 2 c.c. of H_2SO_4, and warm again over water-bath till faintly yellow; now add about 25 c.c. of water and excess of ammonium hydrate, and make up to 100 c.c. in a Nessler glass. The yellow tint produced is due to the formation of ammonium picrate. We next make a standard by taking 5 c.c. of a solution of KNO_3 (·722 gramme to the litre), 1 c.c. of which contains ·0001 gramme of N, and treat with phenylsulphate water, and ammonia as before. We must now compare the tints of the two solutions by looking down through the liquid in the Nessler glasses, holding them about an inch above a sheet of white paper. If the tints are equal, the 50 c.c. of sample taken contains the same amount of nitrogen as the 5 c.c. of KNO_3 solution—i.e., ·5 milligramme, and, therefore, ·5 × 2 = 1·0 part per 100,000. If the sample is not equal to the standard, but is either darker or lighter, pour out from the cylinder which has the darkest tint into a graduated measure until the colours are equal; then calculate as below:

(1) The standard is darker than the sample. Pour out from the standard into a measuring cylinder until the tints are equal. Suppose 25 c.c. have been poured out; then, since the standard contains ·5 milligramme of N, the sample will contain $\frac{75}{100}$ of ·5 = ·375 milligramme of N, and as 50 c.c. of water were taken, this is ·375 milligramme × 2 = 0·75 part of N per 100,000.

(2) The sample is darker than the standard. Pour away from the sample until the tints are equal. Say we pour

away 25 c.c. as in last example; we have 75 c.c. left,
therefore 75 c.c. of sample are equal to the whole of the
standard, i.e., ·5.

Therefore, the whole of the sample will be equal to
$\frac{100}{75} \times ·5 = ·66$ milligramme of N. And this being the
amount in 50 c.c. of the water, there will be ·66 × 2
milligramme in 100 c.c = i.e., 1·32 part per 100,000.

In practice the accuracy of this process will be increased
by taking such a quantity of water as is likely to yield
approximately ·5 of a milligramme of oxidised nitrogen;
thus in the case of a sewage effluent or polluted well-water
a quantity of 20 c.c. might be evaporated, while in an
upland water or an unpolluted river or town-supply a
quantity of 200 c.c. would be better.

The standard nitrate solution is made up so that every
cubic centimetre shall contain ·1 of a milligramme or
·0001 gramme of nitrogen; therefore if we intend to prepare
a litre of solution, we shall require ·1 gramme of nitrogen
in 1000 c.c.

To ascertain how much potassium nitrate (which is the
salt generally employed) we must take, it is necessary to
see how much nitrogen potassium nitrate contains, the
formula being KNO_3, and the molecular weight :

$$(K = 39·1, N = 14, O_3 = 48)\ 101·1,$$

so that 101·1 parts of KNO_3 contain 14 parts of N. If
14 parts of N are contained in 101·1 parts of KNO_3, then
1 part N will be contained in $\frac{101·1}{14}$ parts of $KNO_3 = 7·22$
parts; so that to obtain sufficient KNO_3 to yield ·1 gramme
of nitrogen, as in the solution above mentioned, we shall
require ·722 gramme of potassium nitrate.

Detection of Nitrites.

The two chief tests for the recognition of nitrites are the iodine test and Griess's metaphenylene-diamine test, while very many others have been described, in some of which quantitative results are aimed at. The iodine test may be performed thus: To 20 or 30 c.c. of the water add a few drops of sulphuric acid, a crystal of potassium iodide free from iodine, and a few drops of chloroform. On shaking up and then allowing to stand, so that the scattered globules of chloroform coalesce, it will be found that a very minute trace of nitrites will have coloured the chloroform pink by the liberation of iodine. Instead of the chloroform, starch may be used as an indicator, but in any case especial care must be taken to do a blank experiment on distilled water with the same reagents for comparison, otherwise no safe conclusion can be drawn, as it is difficult to keep potassium iodide without its decomposing, even in the solid state in the dark. The solution, if exposed to light, very soon contains free iodine, and therefore a crystal should be employed. If Nessler glasses containing water to which potassium iodide solution is added are exposed to sunlight, a *few minutes* are sufficient to cause a constant production of free iodine, while even diffused daylight will have the same effect after some hours. Even in the solid state potassium iodide will decompose in the light, and should be kept in a dark cupboard.

Griess's metaphenylene test consists in the addition of a few drops of sulphuric acid to the water to be tested in a Nessler glass, and then a little solution of metaphenylene-diamine hydrochloride.

A minute trace of nitrites results in the production of an orange-yellow colour. The solution should be freshly prepared each time, and a blank should be done on distilled water.

If it is desired to attempt a quantitative estimation of nitrites by comparing the colour produced in a sample with the colour produced in a standard solution of a nitrite, care must be taken to employ a freshly-prepared standard solution of nitrite of potash or soda, which would be best made from silver nitrite, which is a stable salt, whereas the nitrites of potash and soda are very unstable even in the solid form.

It is probable that almost all the compounds of nitrogen that enter a water from any source are eventually oxidized into nitrates, and either remain in that condition or are reduced by oxidizable matter to ammonia, and again re-oxidized to nitrates and taken up by the roots of growing plants.

GASES DISSOLVED IN WATER.

Water readily dissolves or absorbs any gases or vapours coming in contact with it. It is to the presence of dissolved air and carbon dioxide that the pleasant, sharp taste of fresh water is chiefly due. When these gases are absent, or nearly so, water has a flat, almost nauseous taste, as is readily seen in the case of freshly boiled distilled water.

As has been remarked elsewhere, the degree of aeration of water is no criterion of its purity; the most highly polluted water may be fully aerated, and this is particularly the case when the pollution is recent, and therefore probably presents its maximum danger. Water that is fully aerated by prolonged contact with air, such as rain-water or waters that have passed over weirs or waterfalls, contains at the average temperature and pressure about

20·77 c.c. of dissolved gases to a litre, the average percentage composition of which is:

Nitrogen	...	...	63·1
Oxygen	...	...	30·7
Carbon dioxide	...	...	6·1

The amount of gases that a water will dissolve varies with the temperature and pressure. Gases that are moderately soluble in water follow the law of Dalton and Henry—that is to say, the amount of gas dissolved is proportional to the pressure. According to Fownes, one volume of water at a pressure of 30 inches dissolves the following volumes of gases, measured at 0° C. and 30 inches pressure:

Temp.	Oxygen.	Nitrogen.	Carbon Dioxide.	Chlorine.	Sulphuretted Hydrogen.
0° C. ...	0·041	0·020	1·80	—	4·37
10° C. ...	0·033	0·016	1·18	2·59	3·59
20° C. ...	0·028	0·014	0·90	2·16	2·91

These factors have further to be corrected in accordance with the law that the pressure which determines the solubility of a mixture of gases in water is not the general pressure to which the water is exposed, but that pressure which each gas under consideration would exert if it alone were present in the space with which the water is in contact. Thus in an atmosphere of 21 per cent. O and 79 per cent. N the theoretical percentage composition of dissolved air at 0° C. and 30 inches pressure would be 0·21 × 0·041 O and 0·79 × 0·020 N, assuming the atmosphere to be indefinitely large. If, however, the absorption occurred in a partly full bottle, in which the percentage composition of the residual air altered during the process, these proportions would be altered. On this ground the sample bottle must be filled quite full; and on account of the reaction between oxygen and some constituents, living or otherwise, of water, it is of the first importance to accurate results that the determination of oxygen should be made with the least possible delay after the sample has been taken.

The Estimation of Dissolved Gases.

The most simple and satisfactory method of estimation of the dissolved gases in water is by means of an apparatus devised by Mr. S. Harvey (*Analyst*, xix., 121). It consists of a spindle-shaped glass vessel, with two opposite taper-

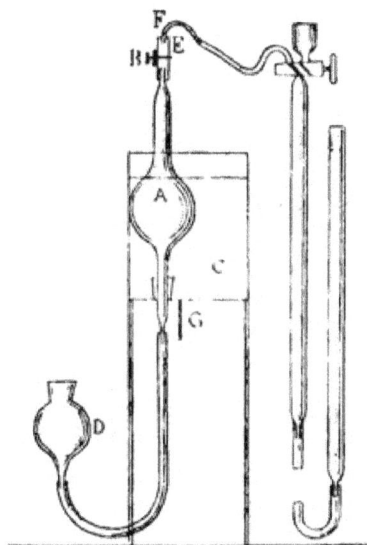

HARVEY'S APPARATUS.
(*Obtainable from Messrs. Townson and Mercer.*)

ing necks, 13 to 14 inches long between extremities, the globular part being 3¼ or 3½ inches in diameter. The upper neck is somewhat enlarged, tapering gradually; the lower is narrow and cylindrical; both end in a capillary bore with a swollen tip, in order that the rubber tubing may be securely affixed. The upper one has a short length of small-bore pressure tubing, securely tied and furnished with a screw-clamp.

The apparatus is used as follows: In the first place, the exact capacity of the vessel from end to end is ascertained, once for all. It is then filled with the sample to be analyzed, the screw-clamp is closed, and the vessel carefully fixed upright in a tin water-bath 6 inches in diameter, and 7 inches high, standing on legs 12 inches high, and having an opening in the bottom closed by a perforated cork, and so arranged that while the lower stem of the spindle projects 2 inches below the bottom, the globular part is immersed in the bath itself.

The clamp is now opened, and about a third of the water allowed to run out into a graduated vessel. Of course the amount deducted from the capacity of the globe when full gives the amount experimented upon. The lower end of the spindle has now 2 feet of small-bore pressure tubing slipped over it, and secured. This tubing has a mercury reservoir at the other end, and the latter suitably supported.

Clean, pure mercury is now poured into the reservoir, the clamp is opened, and the air, together with any bubbles, driven out, the water being allowed to follow to the upper end of the rubber tube. The clamp is now closed again. An ordinary nitrometer, having a bent capillary glass tube affixed to the beak, is now filled with mercury, and the mercury forced to the end of the capillary tube, which is thrust into the top of the rubber tubing and secured. The reservoir is now lowered, and the clamp cautiously opened in order to draw a little mercury sufficiently far to reach the lower end of the capillary tube. The clamp is again closed, the water-bath filled with cold water, and heat applied. To prevent the latter injuring the lower end of the spindle, a metal curtain is riveted on to the bottom of the bath, so as to screen the glass from the flame, and for the same reason the hole in the bath is excentric, to allow sufficient space for heating.

Under the diminished pressure caused by lowering the cistern, the water in the spindle soon boils, and, so far as the author's experience goes, without 'bumping,' and the expelled gases collect in the upper stem. After two hours' boiling, during which the process requires but little attention, the reservoir may be raised, the clamp opened, and the gases passed into the nitrometer, care being taken not to let the 'following water' rise as far as the capillary part of the spindle. The clamp is once more closed, the reservoir lowered, and the operation continued, in order to see if any more gas appears. Finally, the reservoir is raised, and the residual gas driven completely into the nitrometer, the 'following water' being allowed this time to go as far as the nitrometer tap.

The apparatus is now disconnected from the nitrometer, the contents of which latter, after cooling, are subjected to measurement and absorption as usual.

The advantages claimed for the above apparatus are as follows: The water treated, together with the evolved gases, do not quit the vessel until the end of the operation, when the latter are delivered free from splashings into the measuring tube.

There are no corks, perforated or otherwise, and but few connections, while the latter are of a character admitting of being secured against leakage. The apparatus while working requires but little attention; so long as the water in the bath is kept boiling, that in the vessel may be regulated efficiently by due adjustment of the mercury cistern.

The nature of the gases after they have been collected and measured in the above apparatus can be determined by the usual method of absorption.

The carbon dioxide (and sulphuretted hydrogen when present) may be absorbed by caustic potash ; the residual gases, after drying, will consist of oxygen and nitrogen.

The oxygen is absorbed by means of alkaline pyrogallate; the residue, after drying, will be nitrogen. All the gaseous volumes are reduced to standard temperature and pressure.

The Estimation of Carbon Dioxide.

For determining the amount of carbon dioxide which exists both free and as bicarbonates, the method devised by Pettenkofer will be found the most convenient. It is carried out as follows:

To 200 c.c. of the sample are added 10 c.c. of a neutral saturated solution of calcium chloride, 5 c.c. of a saturated solution of ammonium chloride, and 35 c.c. of standardized baryta-water, making in all a volume of 250 c.c. The calcium chloride will decompose any alkaline carbonate, and the ammonium chloride will prevent the precipitation of any magnesia present. The flask is closed with a tightly-fitting rubber stopper, well agitated for some time, and allowed to stand for about twelve hours. Two portions of 50 c.c. are then carefully decanted off, taking care not to disturb the precipitate, and the free, uncombined baryta titrated by means of a standard solution of oxalic acid, using phenol-phthalein as indicator. The standard solution of oxalic acid is so arranged that 1 c.c. = 1 milligramme of CO_2. This is made by dissolving 2·8636 grammes of pure crystallized oxalic acid in a litre. The number of c.c.'s required of standard oxalic acid solution is multiplied by 5, and this number deducted from the number of c.c.'s of standard acid required to neutralize 35 c.c. of the original baryta-water. The difference gives the amount of baryta precipitated by the CO_2 in the water, both free and as bicarbonate.

Example of calculation :

50 c.c. of the clear decanted liquid required 3·0 c.c. of the standard solution of oxalic acid.

$3 \times 5 = 15$ c.c. standard oxalic acid for 200 c.c. sample.

35 c.c. original baryta-water required on titration 31·0 c.c. standard oxalic acid.

$31 - 15 = 16$ c.c. standard oxalic solution.

Since 1 c.c. oxalic acid solution = ·001 mgm. CO_2,

$\therefore$ $16 \times ·001 = ·016$ milligramme CO_2 in 200 c.c. sample taken = ·008 milligrammes per 100 c.c. (equivalent to parts per 100,000).

Free Carbon Dioxide.

Free CO_2 can be approximately estimated, as proposed by Reichardt, by titrating with standard baryta-water, using rosolic acid as indicator. This indicator is prepared of 0·2 per cent. strength in 80 per cent. alcohol, and just neutralized with baryta-water until a permanent red tint is formed. 0·05 c.c. of this reagent is added to 50 c.c. of the sample, and on titration with baryta-water the red tint will ultimately change to yellow. A slight excess is required to produce the change, and the figure returned should be the mean of several determinations.

Estimation of Sulphuretted Hydrogen.

Sulphuretted hydrogen frequently exists in waters from volcanic regions and those derived from certain geological formations, of which the mineral water of Harrogate is a good example in this country. Waters may also be contaminated with sulphuretted hydrogen as the result of the putrefaction of organic refuse, and from contamination with alkali and other factory waste. When sulphuretted hydrogen is present in water to an appreciable extent it may be estimated by the following process :

To 500 c.c. of sample are added 10 c.c. of $\frac{N}{20}$ iodine solution. This is then titrated back with $\frac{N}{20}$ solution of sodium thiosulphate, using freshly-made starch paste as indicator towards the end of the reaction. Each c.c. of the $\frac{N}{20}$ iodine solution used $= 0\cdot0085$ of H_2S. The sulphuretted hydrogen decolorizes the iodine according to the following equation :

$$I_2 + H_2S = 2HI + S.$$

Care must be taken to have an excess of iodine; if there is sufficient sulphuretted hydrogen to decolorize the iodine, more must be added.

Estimation of Dissolved Oxygen.

The estimation of free oxygen in water has a very important bearing, particularly in the case of river waters, when the amount of free oxygen present bears a very distinct relation to the amount of pollution. The results of a valuable investigation made by W. J. Dibdin upon the River Thames show this very clearly. The following table shows the average quantity of dissolved oxygen found by him at various points on the river from July, 1893, to March, 1894 :

AVERAGE PERCENTAGE QUANTITY OF OXYGEN (SATURATION $= 100$)
DISSOLVED IN THE THAMES AT

	High Water.	Low Water.
Teddington (above weir) ...	85·0	
Kew	70·3	85·5
Battersea	42·6	67·6
London Bridge	34·5	51·8
Blackwall	22·5	34·3
Woolwich	22·2	30·8
Crossness	43·0	41·6
Gravesend	50·7	89·5
The Nore	90·1	89·1

A very ingenious, exact, and easily worked process for the determination of free oxygen in water has been devised by Dr. Thresh, and depends upon the following facts. When sulphuric acid and potassium iodide are added to a water containing a nitrite, the amount of iodine liberated varies with the length of time during which it is exposed to the air. If air be excluded there is no increase in the amount of iodine liberated after the first few minutes. If the water is absolutely free from dissolved oxygen, and is kept in an atmosphere of coal-gas or hydrogen, a still smaller amount of iodine is liberated :

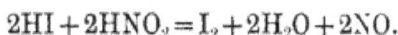

$$2HI + 2HNO_2 = I_2 + 2H_2O + 2NO.$$

When oxygen has access to the solution, the nitric oxide acts as a carrier, and more hydrogen iodide is decomposed, the nitric oxide apparently remaining unaffected, and capable of causing the decomposition of an unlimited quantity of the iodide :

$$2HI + O = H_2O + I_2.$$

This reaction is the one utilized by Dr. Thresh in the process he has devised for estimating the oxygen dissolved in water. The process is described by him (*Journ. Chem. Soc.*, vol. lvii., p. 185) as follows :

The following are the reagents required :

(1) Solution of sodium nitrite and potassium iodide.

Sodium nitrite	... ...	0·5 gramme.
Potassium iodide ...	...	20·0 grammes.
Distilled water	... ...	100 c.c.

(2) Dilute sulphuric acid.

Pure sulphuric acid	... ...	1 part.
Distilled water	... ...	3 parts.

(3) A clear, fresh solution of starch.

(4) A standard solution of sodium thiosulphate.

Pure thiosulphate ...	...	7·75 grammes.
Distilled water	... ...	1 litre.

(1 c.c. corresponds to 0·25 milligramme of oxygen.)

The apparatus required is very simple, and consists of a wide-mouthed white glass bottle of about 500 c.c. capacity, closed with a caoutchouc stopper having four perforations. Through one passes a tube, drawn out at its lower extremity

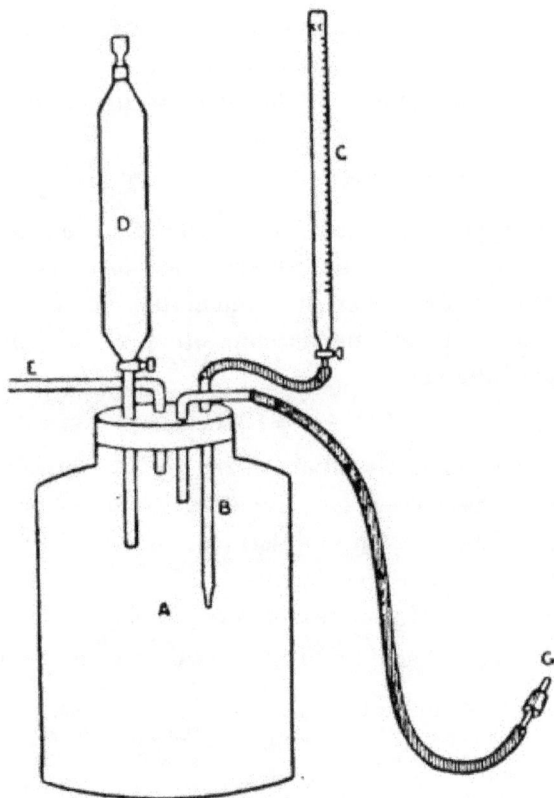

THRESH'S APPARATUS.

to a rather fine point, and connected at the upper end, by means of a few inches of rubber tubing, with the burette containing the thiosulphate. Through another opening passes the nozzle of a separatory tube, having a stopper and stopcock. The capacity of this tube, when full to the stopper, must be accurately determined. Through the third opening passes a tube, which can be attached to the ordinary gas supply. Through the last aperture is passed another tube for the gas exit, and to this is attached a sufficient length of rubber tubing to enable the cork at its end to be placed in the neck of the separator when the stopper is removed. A small piece of glass tube projects through the cork, to allow of the escaping gas being ignited.

The apparatus is used in the following manner: The bottle A being cleaned and dry, the perforated bung is inserted, the burette charged, and the tube fixed in its place. Another tube E is connected with the gas-supply. The separatory tube is filled to the level of the stopper with the water to be examined, 1 c.c. of the solution of sodium nitrite and potassium iodide is added from a 1 c.c. pipette, then 1 c.c. of the dilute acid, and the stopper is instantly fixed in its place, displacing a little of the water, and including no air. If the pipette be held in a vertical position, with its tip just under the surface of the water, both the saline solution and the acid, being much denser than the water, flow in a sharply-defined column to the lower part of the tube, so that an infinitesimally small quantity (if any) is lost in the water, which overflows when the stopper is inserted. The tube is next turned upside down for a few seconds for uniform admixture to take place, and then the nozzle is pushed through the bung of the bottle, and the whole allowed to remain at rest for fifteen minutes, to enable the reaction to become complete. A rapid current of coal-gas is now passed through the bottle, until all the air is displaced and the gas burns at G with a full luminous flame; the flame is now extinguished, the stopper removed from the separator, and the cork rapidly inserted in its place. On turning the stopcock, the water flows into the bottle. The stopcock is turned off, the cork removed, and the supply of gas regulated so that a small flame only is produced when this gas is ignited. Thiosulphate is now run in slowly until the colour of the iodine is nearly discharged. A little solution of starch is then poured into the separatory tube, and about 1 c.c. allowed to flow into the bottle by turning the stopcock. The titration with thiosulphate is then completed. After the discharge of the blue colour, the latter returns faintly in the course of a few seconds, due to the oxygen dissolved in the volumetric solution ; after standing about two minutes, from 0·05 to 0·1 c.c. of thiosulphate must be added to effect the final discharge. The amount of volumetric solution used must now be noted. This will represent a, the oxygen dissolved in the water examined; $+b$, the nitrite in the 1 c.c. of solution used, and the oxygen in the acid and starch solution ; $+c$, the oxygen dissolved in the volumetric solution used. To find the value of a, it is obvious that b and c must be ascertained. This can be effected in many ways, and once known, does not require redetermination unless the conditions are changed.

To Find the Value of b.—Probably the best plan is to complete a determination as above described, and then, by means of the stoppered tube, introduce into the bottle in succession 5 c.c. of nitrite solution, dilute acid, and starch solution. After standing a few minutes, titrate. One-fifth of the thiosulphate used will be the value required.

To Find the Value of c.—This correction is a comparatively small one, and admits of determination with sufficient accuracy, if we assume that the thiosulphate solution normally contains as much dissolved oxygen as distilled water, saturated at the same temperature. Complete a determination as above described, then remove the stoppered tube, and insert a tube similar to that attached to the burette, and drop in from it 10 or 20 c.c. of saturated distilled water exactly as the thiosulphate is dropped in. Allow to stand a few minutes and titrate. One-tenth or one-twentieth of the volumetric solution used, according to the number of c.c. of water added, will represent the correction for each c.c. of volumetric solution used. Call this value d.

Let e be the number of c.c. of thiosulphate used in an actual determination of the amount of oxygen in a sample of water,

f = the capacity in c.c. of the tube employed $- 2$ c.c., the volume of reagents added, and

g = the amount of oxygen in milligrammes dissolved in 1 litre of the water;

Then
$$g = \frac{1000}{4f} (e - b - ed).$$

With a tube made to hold exactly 250 c.c. (the most convenient quantity to use) $\frac{1000}{4f}$ becomes unity, and

$$g = e - b - ed.$$

The following are some results given by Dr. Thresh in his paper :

Source of Water.	f. Amount of Water required.	e. Thiosulphate required.	$e - b - ed$.	g. Milligrammes O, per litre.
1. Spring-water ...	232·5	9·5	7·11	7·64
2. ,, ...	322·0	12·35	9·87	7·66
3. Rain-water ...	232·5	10·15	7·74	8·32
4. ,, ...	322·0	13·05	10·55	8·19
5. Shallow well-water	232·5	8·95	6·57	7·07
6. ,, ,,	322·0	11·35	8·90	6·91
7. Rain-water ...	232·5	9·9	7·49	8·05
8. ,, ...	322·0	12·95	10·45	8·11
9. Distilled-water ...	232·5	12·05	9·58	10·30
10. ,, shaken with air	322·0	16·00	13·40	10·40

7

Consideration of the Results obtained by the Foregoing Processes.

In judging the data obtained as the result of the analysis of a sample of water, all the factors have to be very carefully considered before any opinion can be formed as to whether it has suffered from organic pollution. In giving an opinion upon a water analysis, it cannot be too strongly emphasized that, although the analyst may prove the presence of polluting organic matter, it is impossible in the present state of our knowledge to state positively that such impurity is injurious or innocuous. In doubtful cases a water should never be judged upon one analysis alone, and although the chemical data obtained are to a great extent an index of the amount of pollution, their value can only be properly assessed by a knowledge of the source of the water-supply in question.

It cannot be too clearly recognised that the 'analysis' of water, which aims at discovering that which is not the adulterant or substance to be avoided, but is a probable accompaniment of it, is broadly distinguished by this fact from other chemical analysis, which determines the substance itself. When this fact is mastered, as by the lay public it rarely is, it becomes evident that such process of examination, which can at best weave round the offending substance a chain of circumstantial evidence, requires a vastly larger number of observations for the formation of a sound opinion than does ordinary analysis, which is capable of directly catching the offender red-handed. It may be deplorable that no more direct assistance in the examination of water is to be obtained from chemistry. This does not, however, alter the fact that such assistance, if rightly used, is the best safeguard at present known to science. Every analyst to whom the opportunity occurs is bound to bring these facts to the knowledge of his clients, and to dissuade them from the babyish pretence that chemical analysis can attack water in the same way as it attacks other substances. The duty of a sanitary authority responsible for a water-supply is to keep it under continuous chemical observation; and an analyst should make his client understand the risks which are incurred by being content with the examination of sporadic samples.

Water Standards.

A large number of attempts have been made from time to time by various authorities to lay down a standard to which water for domestic purposes should conform.

The impossibility of the formulation of such a general standard for waters intended for domestic purposes cannot be too greatly insisted upon. They cannot be framed in such a way as to be applicable to districts situated widely apart, unless, indeed, such limits are made so wide as to be partially useless.

Absolute agreement regarding standards is not to be found among the authorities, except, perhaps, as regards the statement that water of the highest purity should be clear, colourless, odourless and tasteless. Very many of the limits given for one constituent are conditional upon the presence or quantity of others, and many investigators maintain that, in judging of the quality of a water, the average of the constituents in water of the same district should be considered. After all, the important matter seems to be to make sure that the water has not undergone contamination from animal sources.

Recent writers on water analysis and hygiene adopt the view that, while general standards are unnecessary and possibly misleading, local standards might be very useful. Professor Nichols (*Water Supply*, 1886) says: 'Moreover, it cannot be insisted upon too strongly that different classes of water cannot be judged by the same standard, and the results of the analysis of waters belonging to different classes ought not to be put into the same table, or otherwise arranged so as to invite comparison. If within the same geological area it is possible to analyze the water from a considerable number of unpolluted wells, a standard may be fixed for the well-water of that region, and a surface-water may be compared with other surface-waters of the same, or of a similarly situated region ; or a stream in one part of its course may be compared with its own unpolluted headwaters.' Professor Mallet (*National Board of Health Bulletin*, 1882) says : 'There seems to be no objection to the establishment of *local* "standards of purity" for drinking-water, based on sufficiently thorough examination of the water-supply in its usual condition.' Dr. Dupré (*Analyst*, vol. v., p. 215) says : ' In the first place, I would caution analysts most strongly against the adoption of any general standards of purity, such as are laid down by some chemists. We may, of course, find waters of such absolute purity that we can at once safely pronounce them fit for all domestic uses; but, short of this highest purity, it is dangerous to rely on any general standards. The only safe standard to go by in any individual case is the standard of purity furnished by unpolluted waters of the district from which the sample under examination has come.' Similar opinions are also held by such high authorities as Adams, Hehner, Thresh, and others.*

The following table of the averages given of the composition of unpolluted waters (mean of 589 analyses) in the

---

* In an inquiry recently made by one of us on some samples of mineral waters submitted by manufacturers to the Medical Press the variations which will occur in waters which are otherwise shown to be pure are well exemplified, and we reprint the figures in an Appendix.

sixth report of the Rivers Pollution Commission of 1868 is inserted because of the manner in which the various samples are classified. The following statement shows the contents in parts per million of certain constituents in such unpolluted well-waters divided into four classes :

	Class I. Rain-water.	Class II. Upland-water.	Class V. Deep Well-water.	Class VI. Spring-water.
Dissolved solids	29·5	96·7	437·8	282·0
Nitrogen, as albuminoid ammonia	0·15	0·32	0·18	0·13
Nitrogen, as free ammonia	0·29	0·02	0·12	0·01
Nitrogen, as nitrates and nitrites	0·03	0·09	4·95	3·83
Chlorine	2·20	11·30	51·10	24·90
Organic carbon	0·70	3·22	0·61	0·56

The usual analytical results from uncontaminated waters, per million in parts, as given by Leffmann and Beam, are shown in the following table :

LEFFMANN AND BEAM'S ANALYTICAL RESULTS (PARTS PER MILLION).

	Rain.	Surface.	Subsoil.	Deep.
Total solids	5 to 20	15 and upwards	30 and upwards	45 and upwards
Chlorine	Traces to 1	1 to 10	2 to 12	Traces to large quantity
Albuminoid ammonia	0·08 to 0·20	0·05 to 0·15	0·05 to 0·10	0·03 to 0·10
Free ammonia ...	0·20 to 0·50	0·00 to 0·03	0·00 to 0·03	Generally high
Nitrogen, as nitrites	None or traces	None	None	None or traces
Nitrogen, as Nitrates	Traces	0·75 to 1·25	1·50 to 5·00	0·00 to 3·00

VI. THE BACTERIOLOGY OF WATER.

General Considerations.

THE space at our disposal does not permit of our giving any general account of bacteria, nor of the methods of staining or cultivating which are employed for their recognition, except in the briefest manner. We have given elsewhere* the principal information, together with descriptions of some of the better known bacteria.

It has already been explained that the chemical substances, the estimation of which has formed the subject of preceding articles, are not necessarily or usually in themselves dangerous, or, so far as we know, even injurious to health. Until recent years, the only reason which existed for using them as indices of the quality of a water were the facts that in waters which had been proved to cause disease they had usually been present in quantities beyond those found in normal waters, and that they were present in large quantities in sewage, which an ancient instinct had condemned as unhealthy for drinking purposes. The development of the science of bacteriology has enabled us in some measure to assign the precise substances which are injurious; and on present knowledge it may be broadly stated that the necessary condition of

* 'Applied Bacteriology.' Second Edition. Baillière, Tindall and Cox.

a water being capable of causing disease is the presence in it of pathogenic bacteria.

At first sight this fact appears to suggest that the science which revealed it has also furnished us with a direct and certain criterion of the potability of a water, and that a bacteriological examination would give us immediate and conclusive evidence as to its wholesomeness. The first point to be noted in the bacteriological examination of water is that at present any such inference would be misleading and unfounded.

All natural waters necessarily contain micro-organisms, as they are constantly being carried into it by air-currents and by the drainage from land-surfaces. It is only in water from deep artesian wells and deep-seated springs that organisms are very few in number, but it is very rare even in these to find them entirely absent. The number and variety of the bacteria in water depends upon several conditions, such as the amount of organic matter in the water, temperature, depth, whether running or stagnant, pollution, source, etc. In stagnant water, such as is found in ditches and small ponds, the number of micro-organisms present is always very great. The comparatively pure water of large lakes and upland streams often contains many bacteria, but these are usually harmless saprophytes, which find their normal habitat in such waters. Water, indeed, forms the most active vehicle for the distribution of bacteria, but the number contained therein varies very much with the source of the water. The bacteria naturally found in unpolluted waters will vary greatly in kind as well as in numbers ; but for the most part they will be chromogenic, non-liquefying, and will develop best at about 22° C., not at blood-heat. Thus, in the purest upland streams and lakes we frequently find that the number of bacteria in 1 c.c. is under 100, while in town sewage there are many millions in the same volume. In ordinary rivers the number is generally between 1,000 and 100,000 per c.c. In the case of waters from deep-seated springs, the presence of more than 100 organisms per c.c. is fairly conclusive evidence that the water has undergone some contamination with surface-water. Micro-organisms are also found, although not in great numbers, in rain-water, hail, snow, and even in the ice of glaciers.

Now, it happens with bacteria, as with every other living thing, that the circumstances in which any organism normally occurs are those which are most favourable to its development; and, contrariwise, an organism in abnormal circumstances has usually a poor chance of survival. The normal habitat of organisms of disease is not water. The extraordinary development which they show in the bodies of susceptible persons under conditions of nourishment and temperature which never occur in water shows how unlikely it is that the same organisms should thrive in the entirely different conditions found in water. It is, accordingly, the fact that the bacteria which occur among the normal flora of water are not pathogenic, and the first difficulty which arises in applying bacteriological examination to water is to determine the account which is to be taken of the harmless water bacteria.

In the early stages of the science it was proposed simply to count all the bacteria in water, and on the ground that many waters known to be pure contained not more than 100 bacteria per c.c., to regard as wholesome any water which did not contain more than that number. In considering such a criterion, a careful distinction must be drawn between a standard for a water which has passed through a purification process, such as filtration, and that for a natural water. In each case bacteriology has shown that by far the most important source of infection, and indeed the only one of which in ordinary circumstances account is taken is sewage. With a natural water it is, as we have seen, difficult to say with any certainty that the possibility of sewage pollution is absolutely excluded. It is, however, possible to do so in some circumstances ; for example, in water from inaccessible mountains before it has reached the level of habitations. In such cases it is the local circumstances which furnish the evidence of safety, and the number of bacteria present is absolutely immaterial, except in so far as it indicates a nutrient value in the water which might favour the development of any infection by which the water was subsequently affected. Accordingly for such waters it might be safe to say that if they contain less than the assumed standard for pure waters, the poorness in microbes confirms the inference of purity derived from other sources of information ; if they contain more, the fact would in such a water be immaterial. On the other hand, when the water in question is one which has undergone a process of purification, the case is entirely different. It is no longer that of a water which by local conditions is protected from pollution, and can therefore only contain microbes of a harmless character. It is the case of a water which required, or by its local circumstances was liable to require, purification ; and it is to be observed that this is the case of the large majority of water-supplies in this country. The presence in such a water of a number of bacteria lower than the standard is no longer reduced to the presence of harmless organisms by process of geological and similar exhaustion which was applied in the former case. It means merely the removal of a proportion of a larger number of organisms, of which some were, or were liable to have been, pathogenic ; and with the ordinary processes there is nothing to show that the organisms which find their way into the filtrate are harmless. On the doctrine of probabilities they will be so more often than not; but with equal certainty the same doctrine teaches us that in the long-run they will at some time prove to be harmful. For the rare cases where purification is unnecessary, conformity with a standard number of organisms may be a satisfactory test, but where purification has been required, it is utterly impossible to fix any standard at all; and the presence of bacteria in the filtered water is a sign, and their number is to some extent a measure, of the deficiency of the purifying process.

What, then, is the use of the numeration of bacteria? The answer is that in the present state of knowledge such examination is a process of precisely the same significance as the chemical processes which we have already discussed.

The number of bacteria in a stream or river will evidently vary very considerably according to the temperature—in summer there will be more than in winter, according to the rate of flow; if the flow is very slow, the bacteria will fall to the bottom, partly because they are of greater specific gravity than water, and partly because they are carried by still denser particles. If water comes straight from springs, or from uncultivated uplands on which vegetation is sparse, the bacteria will be few; while if it flows through or from cultivated and inhabited land the number will be greater. After a fall of rain we should expect, not a decrease, but a considerable increase in the number and species of bacteria present in a stream or river; owing to the fact that a great number of bacteria will be washed off the banks into the stream, and that a great number which were previously present in the stream, but had settled to the bottom, had again been stirred up owing to the increased velocity of the current.

The variations due to temperature can to some extent be allowed for; those due to rainfall may be tolerated when the source is above habitation level. But other variations denote the influx of some matter not normally in the stream, and the main value of counting bacteria lies in the fact that any such variation must be regarded as pollution, and can often be more readily detected in this way than by chemical analysis. This is a far less satisfactory result to obtain from bacteriological examination than one would have hoped, but it is still a result which may have a considerable value. The variations may be those of a stream or well examined daily or weekly, or they may be those simultaneously existing in feeders of a single stream all situated in the same geological district, as has recently been suggested by Professor Delépine. Whether in time or place, any large variation in the number of bacteria in a given supply or its constituent sources is an indication which can be readily recognised, and may give grounds for suspicion which might not otherwise or so readily be revealed.

Collection of Samples for Bacteriological Examination.

It has been already pointed out that bacteria are heavier than water, and will therefore sink to the bottom to a large extent if the water is at rest. Hence we are at the outset confronted with a grave difficulty which is not sufficiently realized or provided against.

It is too often taken for granted that bacteria are evenly distributed throughout a mass of water as though they were in solution, which is not the case.

In taking a sample of water from a tap or pump, the water should, therefore, be allowed to run for some minutes, while in the case of a river, lake, or reservoir, it can only be possible to get a representative sample by taking several different samples at different depths and different distances from the banks. In any case, the greatest care must be taken not to stir up mud, which is very rich in bacteria, and, on the other hand, to avoid floating scum. In routine work, where, as should always be the case, examination is made at short, periodical intervals, the sample should be taken always at a standard depth from the surface and at a given spot, and the collection of average samples is unnecessary.

Among bacteriologists there is considerable difference of opinion in the amount recommended to be taken as a sample. Unless we are making search for a specific bacillus, *e.g.*, typhoid, the quantity actually *used* in the examination will probably not exceed a few cubic centimetres, and therefore many observers are content to collect 100 to 200 c.c. in glass bulbs. Within limits, however, the larger the sample the better, and we prefer to adopt the recommendation of Klein, which is to collect the sample in an ordinary 'Winchester,' which has been previously washed out with a few c.c. of strong sulphuric acid and then thoroughly rinsed out several times on the spot with the actual water to be tested before taking the sample.

The bottle is filled by immersing *completely* in water, and if care is taken to avoid unnecessary touching of the neck or stopper with the hands, it is not likely that the sample will be contaminated.

When the bottle is filled the stopper may be tied down with a piece of oiled silk which has been well washed in the same water, and the bottle should be packed in ice and sent to the laboratory without delay.

If it is preferred to take samples in small glass bulbs, they may be conveniently blown in a tube, and after steam has been passed through a series of them to ensure complete sterilization, the ends of the tubes are sealed off, so that when the steam condenses a vacuum will be left inside. The sealed end of the tube is drawn out to a fine point. This fine point is broken with a pair of sterile forceps under the surface of the water, and after the latter has rushed in and filled the vacuum the bulb is sealed up again with the aid of a spirit-lamp. An objection to the use of such bulbs is the difficulty of getting the water out of them afterwards.

If it is desired to take samples from definite depths, there are various forms of apparatus figured in the text-books for carrying such bulbs and breaking the sealed end when the apparatus has been lowered to the required depth; or a Winchester might be weighted and sunk, and the stopper withdrawn as desired.

If the sample can be examined on the spot, all fear of multiplication of the bacteria on the way to the laboratory is avoided, and the troublesome work of packing a large bottle in ice, so that it will not be loose when the ice melts, is avoided.

In cases when a sample of 200 to 300 c.c. is considered sufficient a stoppered bottle of about this capacity may be used, and this, with sufficient ice, will readily fit into a tin biscuit-box, which, again, is placed in a wooden box holding sawdust to absorb the water from the melting ice.

It is not so necessary to employ ice in winter weather, when the temperature is low, though it would always be safer to do so, bearing in mind the great increase in the number of organisms which takes place on keeping the samples for a short time. Frankland states that a pure water containing, say, 5 organisms per c.c. when freshly drawn, may, even if kept in a sterile flask free from aerial contamination, contain after a few days perhaps 500,000 in the same volume—or, in other words, nearly as many as are found in slightly-diluted sewage.

This increase is due mainly to the fact that when the sample is taken it is sure to be kept in a warmer place than the well, etc., from which it came. Even in samples placed in ice some multiplication may occur. There is no fear of killing any bacteria by placing the samples in ice; none of those that we wish to look for are injured by exposure to the freezing-point even for some days.

In the case of all samples taken for bacteriological examination, an account should be sent with the sample (as in the case of samples for chemical examination), giving the reason for sending the sample, the time of collection, and some description of the source, as well as the name of the person taking the sample. From the foregoing remarks it will be recognised that greater care must be employed in taking samples for a bacteriological examination than is necessary in the case where chemical analysis only is required, although in both cases it is easy by neglecting precautions to introduce errors which would entirely vitiate the results of any subsequent examination.

The Examination of Samples.

The procedure in the examination of samples must be varied according to the character of the sample and the nature of the information we desire to obtain.

At the present time it is impossible to attempt, at all events as a routine procedure, the identification and enumeration of all the bacteria occurring in an ordinary sample of water, and hence we are compelled to adopt some procedure which shall give us as much information as possible, and shall not occupy more time than can be devoted to individual samples.

The difficulties attending on the attempt to identify individual bacteria are very great.

In the first place, all the bacteria are extremely minute, and, being colourless, they cannot be distinctly seen except when examined in stained preparations with the aid of a powerful microscope capable of giving a magnification of at least 700 or 800 diameters.

Then, again, though we may by these means at once recognise varying forms such as bacilli, spirilla, or cocci, and see that some are motile and others are not, and that some are stained by particular staining reagents and the like, these methods of distinction will only serve as means of grouping of a very general character, and will not enable us to fix the identity of any particular organism.

As a preliminary to the attempt to establish the identity of a particular organism, we must examine it in a number of different ways, of which the following are some:

(1) Characters of culture on gelatine, *i.e.*, colour of colony, whether liquefying or not, raised or flat, etc.

(2) Shape of organism.

(3) Motility.

(4) Production of gas, acid, indol.

It is therefore clear that one of the first steps is to obtain cultures—pure, if possible—of as many of the bacteria present in our sample as we can.

By using Koch's method of gelatine plates we can obtain pure cultures of a considerable number of organisms at the expense of very little time and trouble, but this method only deals with one class of organisms out of four, all of which will probably be present in varying numbers, namely

(1) Ordinary aerobes which grow on gelatine.

(2) Anaerobes which will not appear on the gelatine plate.

(3) Organisms that need blood-heat for their development.

(4) Nitrifying bacteria which require special media.

The first class (aerobes growing at 22° C.) are provided for by the gelatine plates. The second, or anaerobes, may be grown by any of the various methods by which oxygen is excluded. The third group, or those requiring blood-heat, can be studied by means of agar-agar plates, while the fourth class, the nitrifying bacteria, can only be grown on media nearly free from organic matter, such as silica jelly.

In reading a report that a certain water contained 'so many bacteria per c.c.,' the figure usually means those that grow on gelatine only—a distinction that must be borne in mind. Even with this reservation, enormous variations of result occur, not only through slight differences in composition of media, but also through the varying time and method of keeping the cultures and counting the colonies. Some bacteriologists keep plates at 'ordinary room temperature,' and their results are naturally apt to vary from those obtained when the cultures are kept—as, to have any scientific or even practical value, they must be—in an incubator at a definite temperature. Some count colonies at the end of twenty-four hours, some later, and the difference in the number developed in an extra day's growth may often be very large. The limit of time for which plates can be kept is fixed by the liquefying bacteria, but as a rule plates should be counted at least after twenty-four, forty-eight, and seventy-two hours. The number of colonies visible to the naked eye is less than those to be seen under the lens, and the lens, again, does not show all that can be seen under the microscope. The accuracy of counting where large numbers are concerned is also to some extent a matter of personal equation. The important thing is, after all, the ratios occurring in successive samples from the same source of the numbers of bacteria, rather than the precise actual numbers; and it is therefore imperative that any routine examination should follow invariable lines as to nature of medium, time and temperature of incubation, and mode of counting.

It becomes necessary therefore to decide on some scheme

for the routine examination of samples; and it is not surprising to find wide divergence in the features which are recommended by various authors for routine examination. Most authorities concur in including the numeration of colonies on gelatine plates; though some divergence occurs as to the choice between a room temperature and a fixed incubator temperature, and no definite agreement has, unfortunately, been arrived at as to the precise composition of the nutrient medium. Importance is attached by some to the ratio between the number of liquefying to that of non-liquefying organisms developed in such examination; others press the differentiation still further, and insist on the precise recognition of the dominant species. Others, making a similar broad distinction, recommend the numeration of organisms developing at blood-heat. Miquel has suggested the inoculation into animals of a peptone culture seeded with the water, with the object of determining its virulence under those conditions. Some authors attach importance to a microscopical examination of a drop of the water; many authorities demand a special examination in all cases for specific pathogenic organisms; and almost all concur in demanding an examination for the presence, and in particular for the abundant presence, of *Bacillus coli*, an organism which, though disseminated to some extent by dirt, is imported into water in large quantities only by sewage. Klein has produced a considerable body of evidence showing that his very characteristic *B. enteritidis sporogenes* is diagnostic of sewage, though doubtless the diagnosis would apply only when it was present in substantial numbers, too large to have been likely to arise from dust. It is, however, to be observed that both these organisms have their normal habitation in sewage; and even when they occur in water in small numbers, and the inference is possible that it is through

8

dust that they have reached it, there is merely one extra link in the chain connecting the water with alvine excreta.

In selecting a scheme for routine examination it is usually impossible for the analyst to include all the factors which may be notable, and it should be recognised, once for all, that, while some practice in bacteriological methods and manipulations may enable an analyst to make the simpler determinations with an accuracy which should be at least as great as that of a bacteriologist proper, and unfailingly to detect some of the most important danger signals, those determinations which depend on the recognition of individual organisms cannot be properly performed without a very long experience of the work and an intimate acquaintance with the organisms themselves. This caution is the more necessary because even to men constantly engaged on such work these determinations are encumbered with difficulties and ambiguities; and in all parts of the world research is proceeding to throw more light on many points, such as the intimate structure of bacteria, which must furnish an increasing number of data for the service of the bacteriologist in the identification of specific organisms. An analyst whose practice does not involve the constant study of the organisms which are in question in the various phases of their life history can do no worse service to himself or his clients than to undertake work which includes their identification. His proper function, which is not the less valuable because it is more modest than that which by an error of judgment he might attempt to subserve, is to chart those variations in which lies the main information afforded by present-day bacteriology as to the purity of a water, and to detect those danger signals which throw suspicion on the supply, and call, if further bacteriological data are required, for the experience of a bacteriologist. At the present time we know in advance that the information which further bacteriological examination can afford is much more likely to deal in probabilities than to demonstrate any certainty in connection with the supply. The analyst's indication of danger should therefore be the starting-point for immediate examination of possible sources of pollution ; and in the present state of knowledge there is no more valuable service which can be rendered to the public health. It may be observed incidentally that the training of the chemist is peculiarly suited to qualify him to conduct those operations which in our judgment fall within his proper domain with an accuracy which may sometimes be wanting in those who approach bacteriology from the purely pathological or biological standpoint. There is reason to believe that mere looseness in the routine operations of bacteriology—such, for instance, as the exact preparation of media—has led to a deplorable number of needlessly conflicting results. However little the chemist may undertake in that science, what he does should therefore be done with the minutest accuracy ; and we believe that the result would be sets of figures far more comparable than many of those which are at present available, and an employment of the elementary utensils of the science to better purpose than has been usual.

We should recommend the following routine procedure as generally useful :

(1) Estimate the number of organisms on gelatine plates at 22° C., both liquefying and otherwise.

(2) Incubate a small portion of the water for twenty-four hours at 37° C., and make other gelatine plates.

(3) Ascertain whether any growth occurs on a phenol and an Elsner plate.

(4) Make an anaerobic culture of the sample-water ; heat

another portion for ten to fifteen minutes to 80° C., and incubate at 37° C. anaerobically in milk.

Particular attention is directed to the remarks, p. 137, on the recognition of specific organisms.

We shall proceed to describe the execution of these estimations in order.

Determination of the Number of Bacteria.

From 0·02 to 0·5 c.c. of the sample, the number dependent upon its purity, is withdrawn by means of a small sterile graduated pipette, and added to a tube containing melted sterile nutrient gelatine at a temperature of about 27° C. The cotton-wool plug of the tube is then replaced, and the contents of the tube gently agitated so as to thoroughly mix the contents without letting bubbles form. Not less than half a dozen plates (Petri dishes) should be made in this way. In the case of an unknown water it is best to inoculate various quantities within the limits named. The whole value of the examination depends on the assumption that each organism, when it is developed sufficiently to be visible, occupies a colony separate from that of its neighbour, and that the number of the organisms in the small quantity under examination fairly represents that of the bulk. It is therefore essential to the utility of the process that the plate shall contain neither too many organisms nor too few, and the process here recommended is the most convenient way to obtain this result. A good plan is to aim at having about 100 colonies on a plate; and a rough preliminary idea may, after a little practice, be had by drying a drop on a slide, staining with gentian violet, and counting the number of organisms. After the tube has been shaken, the plug is again withdrawn, and the contents of the tube poured on to a sterile glass plate or into a Petri dish. Agar melts at a higher point than the death-point of many bacteria, but fortunately does not set till a much lower temperature. In making agar plates the tube, after its contents are melted, is allowed to cool to just above 40° C. ; the sample of water is then added, and the contents of the tube rapidly poured on to the plate, which should have been warmed, say, in the incubator, to prevent the agar from setting in lumps.

As attempting to measure a volume of less than 0·1 c.c. is not a satisfactory operation, it is best, in the case of a water suspected to contain a large number of bacteria, to dilute the water 50 or 100 or more times, as follows, before proceeding to the examination. In the case of sewage or sewage effluents, it is necessary to dilute from 1,000—10,000 times. Small sterile flasks containing exactly 49 c.c. of sterile distilled or, better, sterile natural, water, receive 1 c.c. of the water under examination. This is well mixed, and then 1 c.c. of this first attenuation is taken, and introduced into another flask until the degree of dilution is considered sufficient. Plates are then made from 1 c.c. of each of the various attenuations. The plates are now allowed to stand at a temperature of about 22° C., and examined daily. A considerable number of colonies will be visible under a low power of a microscope in less than two days, though in this time not all the colonies will be visible that may subsequently make their appearance. If we do not use a microscope, but depend on the naked eye, or a lens only, it is best to count the colonies on the third or fourth day from starting the plate, but no hard and fast rule can be laid down as to this. The presence of liquefying colonies often causes the liquefaction of the plate before one is sure that all of the slower-growing colonies have developed, so that the counting should be repeated daily as long as it is practicable. In the case of agar plates, which are not liquefied, counting should be continued until fresh colonies cease to be developed, though as a rule this would have occurred by the fourth day.

The accuracy of the results obtained depends to a very large extent upon the care with which the organisms are distributed through the nutrient medium. Care should also be taken that the original sample of water is well and thoroughly shaken, to distribute evenly the organisms contained therein, before withdrawing the quantity for the examination. The gelatine medium must be prepared *strictly* in accordance with one established receipt, though even with the greatest care it is occasionally impossible to avoid slight differences in composition in the finished product. On this account a sufficient quantity of nutrient gelatine should be prepared before commencing a piece of work, so that the same batch of media may be used throughout. It is desirable to have in particular a strict standard of alkalinity. At the experiment station, Lawrence, Mass., the practice is to use medium which requires 18 c.c. normal alkali to make it alkaline to phenolphthalein after it has been freshly boiled.

The organisms may be counted by means of Wolffhügel's apparatus. This consists essentially of a glass plate divided into squares, each a centimetre square. Some of these squares are subdivided. The plate or dish is laid under this scale, and the number of organisms present is found by counting the number of colonies in a few of the squares; an average is then taken, and the number of organisms present thus calculated. With a little practice, very close approximations are to be obtained with this apparatus.

The following procedure is recommended in the notification issued by the Imperial German Health Department in regard to the filtration of surface waters used for public supplies :

'In order to secure uniformity of method, the following is recommended as the standard method for bacterial examination :

'The nutrient medium consists of 10 per cent. meat extract gelatine with peptone, 10 c.c. of which is used for each experiment.

'Two samples of the water under examination are to be taken, one of 1 c.c. and one of $\frac{1}{2}$ c.c. The gelatine is melted at a temperature of 30° to 35° C., and mixed with the water as thoroughly as possible in the test-tube by tipping backwards and forwards, and is then poured upon a sterile glass plate. The plates are put under a bell-jar, which stands upon a piece of blotting-paper saturated with water, and in a room in which the temperature is about 20° C.

'The resulting colonies are counted after forty-eight hours, and with the aid of a lens.

'If the temperature of the room in which the plates are kept is lower than the above, the development of the colonies is slower, and the counting must be correspondingly postponed.

'If the number of colonies in 1 c.c. of the water is greater than about 100, the counting must be done with the help of the Wolffhügel apparatus.

'In those cases where there are no previous records showing the possibilities of the works and the influence of the local conditions, especially the character of the raw water, and until such information is obtained it is to be taken as the rule that a satisfactory filtration shall never yield an effluent with more than about 100 bacteria per cubic centimetre.

'The filtrate must be as clear as possible, and in regard to colour, taste, temperature, and chemical composition must be no worse than the raw water.

'To allow of a complete and constant control of the bacterial efficiency of filtration, the filtrate from each single filter must be examined daily. Any sudden increase in the number of bacteria should cause a suspicion of some unusual disturbance in the filter, and should make the superintendent more attentive to the possible causes of it.'

Every city in the German empire using sand-filtered water is required to make a quarterly report of its working results, especially of the bacterial character of the water before and after filtration, to the Imperial Board of Health.

Incubation of Original Sample.

It has been stated above that the bacteria found in most natural, unpolluted waters multiply rapidly if the sample is kept at room temperature for a day or two before examining.

If the sample is kept at a temperature of 37° C., different results are obtained—namely, that, speaking generally, in the case of pure waters, no increase will be found to have taken place, and in many cases there is a decided *decrease*, while in sewage-polluted waters a very great *increase* will be observed.

All the organisms normally present in fæces grow and multiply vigorously at blood-heat, whereas the common water bacteria, which grow readily at 22° C., do *not* grow well at 37° C. Hence in an unpolluted water (which may possibly contain many bacteria that develop on gelatine), if we examine it by means of an agar plate, we should find only comparatively few organisms.

The following results of experiments made by one of us show the value of these two methods :

	Polluted surface-well waters.				Waters of average quality.		
	(a) 800	(b) 1,050	(c) 1,400	(d) 10,000	(e) 180	(f) 270	(g) 460
Approximate number of organisms per c.c. in the original water, as determined by a gelatine plate culture	800	1,050	1,400	10,000	180	270	460
Number of organisms per c.c. appearing on an agar-agar plate culture, after incubating at blood-heat for twenty-four hours	220	180	350	5,500	10	5	150
Approximate number of organisms per c.c., after incubating the water at blood-heat, the organisms then being determined by an agar-agar plate	800,000	Exceeded 1,000,000	Exceeded 1,000,000	Exceeded 2,000,000	—	—	400

Pathogenic Bacteria in Water.

The greater number of the pathogenic bacteria do not flourish in water, as they require blood-heat and special conditions for their development.

Practically the only pathogenic bacteria which are at all largely conveyed by water are those of typhoid and cholera, and probably certain others which are the cause of epidemic diarrhœa.

The Isolation of the Cholera Bacillus from Water.—The detection of Koch's comma bacillus (*Spirillum choleræ Asia-*

ticæ) in water, as in the case of the typhoid bacillus, is a matter of some difficulty, as this organism is rapidly overgrown by the ordinary water bacteria. In the examination of suspected water-supplies, the best method to employ for the detection of this organism is to take advantage of the fact first noted by Dunham, that the cholera spirillum multiplies with great rapidity in alkaline saline peptone solution. The suspected water is examined as follows : To 100 c.c. of the water are added 1 gramme each of pure peptone and common salt; the mixture is made faintly alkaline with sodium carbonate, and then incubated at 37° C. At intervals of ten, fifteen, and twenty hours respectively, cover-glass preparations are prepared from the top of the liquid; these are then examined microscopically; if the organisms all have the appearance of the cholera organism, they are at once subcultured into broth and other media. If other organisms (*e.g.*, *B. coli*, which equally gives indol reaction) are present, plates and pure sub-cultures are made from them. The cultures after incubation are tested for the indol reaction ; *i.e.*, the production of a pink colour on the addition of a few drops of pure sulphuric acid.*

It is well known that many impure, especially sewage-contaminated waters, contain spirilla and comma-shaped bacteria, many of which strongly resemble the cholera organism in many ways ; care must therefore be taken that none of these are mistaken for the true cholera organism. None of these spirillum forms, however, give the indol reaction, and Koch is of opinion that the presence of the cholera bacillus in the water is proved if comma-shaped organisms are found which exhibit the indol reaction, and which give rise to the characteristic symptons on inoculation into the peritoneum of guinea-pigs.

* 'Applied Bacteriology,' Second Edition, pp. 79 and 80, and pp. 403 and 404.

The Isolation of the Typhoid Bacillus from Water.—The general procedures adopted for the isolation of the typhoid bacillus have depended chiefly on attempts to devise media on which the typhoid bacillus alone would grow, or that would at least prove unfavourable to the majority of the common organisms, particularly those that liquefy gelatine, so that the typhoid bacillus might grow without being choked out by other organisms present in greater numbers.

As only a few typhoid bacilli might be present, Klein has adopted the plan of filtering a large volume of water, so as to concentrate its bacterial contents. He passes 1,000 to 3,000 c.c. or more of the sample through a small sterile Pasteur-Chamberland filter. By this treatment all the bacteria in the water are retained on the outer surface of the filter. The particulate matter thus retained is then brushed off the outer coating of the filter with a sterile brush into about 10 c.c. of sterile distilled water. One c.c. of this concentration, which contains the particulate matter representing from 50 to 150 c.c. of the original water, is then immediately submitted to plate culture by one of the undermentioned methods, to isolate the colon bacillus and also the *B. typhosus*, if present.

The difficulties attendant on the isolation of this organism are very considerable, and whereas an immense number of waters known to have conveyed typhoid have been examined by competent observers, the cases in which the discovery of the bacillus of typhoid has been reported are very few, and of these a considerable proportion must be accepted with caution as having been reported before we were in possession of sufficiently stringent tests to properly identify the organism and differentiate it from the large group of similar organisms. The distinctions which we now possess form a formidable list, and whether, even now, they really form a *certain* means of identification, cannot be settled without further experience.

Another mechanical method intended to assist in separating the typhoid bacillus from other bacteria, which has been devised by Delépine, depends on the active motility of the organism. He employs two test-tubes, and places one inside the other. The smaller internal one has a hole at the bottom, and the material to be tested is placed in the larger tube, the intention being to get the typhoid bacteria, through their considerable motility, to pass through the hole and so to reach the nutrient broth in the inner tube, while many of the common bacteria will remain outside.

Of the various methods in which reagents are added to media to restrain the growth of the common organism, many modifications have been suggested, and the following are the chief :

(1) **Inhibition by Means of Phenol.**—The *B. typhosus* and the *B. coli communis* are among the limited number of micro-organisms which will grow in the presence of small quantities of phenol, which addition retards or inhibits the common water bacteria, such as the *B. fluorescens liquefaciens*, *B. mesentericus*, etc., the presence of which would liquefy the gelatine, and by their rapid growth would annihilate the *B. typhosus*, if present. The presence of a small quantity of phenol does not in any way interfere with the growth of the *B. typhosus* or the *B. coli communis*, but exhibits a marked inhibitory effect upon the common water bacteria, and by the retardation and suppression of these, the colonies of the *B. typhosus* and the *B. coli communis* have a chance and leisure to appear.

The use of phenol for this purpose appears to be due, in the first instance, to Chantemesse and Widal (*Gazette des Hopitaux*, 1887, p. 202), who used nutrient gelatine containing 0·25 per cent. of phenol. Thoinot (*ibid.*, p. 384), a little later, inhibited the growth of organisms other than the typhoid and colon bacilli, by adding 0·25 per cent. of phenol to the water under examination, which was then incubated at blood-heat and afterwards plate-cultured.

As pointed out by Holz, and confirmed by Dunbar, the above authors use a percentage of phenol which altogether prevents the growth of the *B. typhosus.* Dunbar states that 0·12 per cent. of phenol greatly interferes with the growth of the typhoid bacillus, while in the presence of 0·14 per cent. it will not develop at all. He further states that in the presence of small quantities of phenol the colon bacillus presents stronger resemblances to the typhoid bacillus than usual.

To ascertain if the resisting power of cultures of the *B. typhosus* to phenol differed, we tried the following series of experiments on different cultures of the organism, using varying percentages of phenol, with the following results :

Percentage of Phenol.

	0·05	0·10	0·20	0·30
B. typhosus (*a*) ...	+ ...	− ...	− ...	−
„ (*b*) ...	+ ...	+ ...	− ...	−
„ (*c*) ...	+ ...	+ ...	+ ...	−
„ (*d*) ...	+ ...	+ ...	− ...	−
B. coli communis ...	+ ...	+ ...	+ ...	+

Thus, it is seen that the resisting power of the *B. typhosus* to phenol varies with different cultures. The sample marked (*a*), which was freshly isolated from the dejecta from a typhoid case, had less resisting power than other samples which had been subcultured through many generations.

(2) **Parietti's Method.**—Parietti proposed the use of broth containing both phenol and hydrochloric acid to eliminate the common water organisms. He takes advantage of the fact that the typhoid and colon bacillus will grow in a

slightly acid medium, whereas the majority of other organisms will not.

Parietti's method is as follows: The following solution is prepared: Five grammes of phenol and four grammes of pure hydrochloric acid are added to 100 c.c. of distilled water. From 0·1 to 0·3 c.c. of this solution is added to a series of test-tubes containing 10 c.c. of sterile nutrient broth (= 0·05 to 0·15 per cent. of phenol). From 0·1 to 0·5 of a c.c. of the water under examination (or preferably of the concentrate obtained by Klein's method) is then added to the tubes, the contents are well mixed, and the tubes again returned to the incubator. If, after twenty-four hours' incubation at blood-heat, any of the tubes appear to be turbid, they are submitted to ordinary plate cultivation, and the resulting colonies carefully examined in subcultures. Frankland states that when only a few typhoid bacilli are present, the incubation must be prolonged for forty-eight or even seventy-two hours.

The great objection to the use of phenolated broth is that when cultivated at blood-heat the colon bacillus and its allies multiply at from two to five times as quickly as the typhoid bacillus, even if the latter is not suppressed altogether. This objection also applies, but in a less degree, to phenolated plates, the surface of which may be covered by the expanded colonies of the colon bacillus.

It must be remembered that the preliminary concentration does not improve matters numerically, and, as has been pointed out by Stoddart, it is not unreasonable to suppose that the somewhat violent treatment may have a more injurious action upon the typhoid bacilli than upon the hardier forms. Since by no method at present known can the ratio of the typhoid organism to the colon bacillus and the common water saprophytes be increased, the general tendency being for them to decrease, the difficulty is best

met by increasing the area of the plates rather than by
any method of concentration.   This numerical difficulty
has perhaps not been fully appreciated by many workers
on this subject, but it has been forcibly presented by Laws
and Andrewes in their recent report (*Report upon the
Micro - organisms of Sewage*, presented to the London
County Council, 1894).   In the case, for instance, of a
moderately polluted effluent containing 50,000 microbes
per cubic centimetre, possibly 90 per cent. of these may be
suppressed by the addition of phenol, leaving 5,000 to be
dealt with by plating out.   There would obviously be no
advantage in concentrating such a water, since it is impossible to deal satisfactorily with a plate of ordinary size
containing 1,000 colonies.   It would then be necessary to
subculture every one of these colonies, for the naked-eye
appearances are not to be relied upon.

In practice, however, we prefer to use simple carbol-
gelatine containing 0·05 per cent. of phenol.   This quantity
is quite sufficient to restrain the growth of liquefying
organisms, and, moreover, with this quantity there is less
danger of losing the typhoid bacillus if it is present.

(3) *Elsner's Method.*—Dr. Elsner, of Berlin, has recently
published (*Zeitschr. f. Hyg.*, xxi. 1) the results of an
investigation made to ascertain the possibility of an early
recognition of enteric fever by the bacteriological examination of the stools.   He has been able to recognise the
Eberth-Gaffky bacillus in some cases in as short a time as
forty-eight hours.   Dr. Elsner went over the existing
methods for the separation of the *B. typhosus* and *coli*, with
no better results than have previously been obtained.   In
all cases but one he found that either persistent organisms
other than those sought to be isolated would grow to a
sufficient extent to spoil the plate, or else the *B. coli* would
develop to an extent capable of preventing the recognition

of the typhoid bacillus. The exception was slightly acid potato-gelatine, containing 1 per cent. of iodide of potassium. The process recommended is to boil potato decoction (500 grammes to 1 litre of water) with 10 per cent. of gelatine. Sufficient of a 2 per cent. solution of sodium hydrate is added till only a faint acidity remains, litmus being used as indicator.

Elsner found that the *B. proteus* and *ramosus*, which always grow on carbolized gelatine, either never occurred on this medium, or were rapidly overgrown by the colon bacillus. The *B. coli* grew in twenty-four hours, presenting the usual appearance of that organism on acid media ; the *B. typhosus* was scarcely visible in twenty-four hours, but in forty-eight hours appeared in small, shining, very finely-granulated colonies like little drops of water, which contrasted strongly with the larger coarsely-granulated brownish colonies of the colon bacillus. The *B. coli* only acquired the appearance of the typhoid colonies when a great number of the organisms were present, and many, therefore, grew without finding room for their proper development. In plates made with weaker inoculations it is impossible to mistake one bacillus for the other.

We have used this method with satisfactory results on waters artificially contaminated with the typhoid bacillus. The colonies of the *B. typhosus* appear more quickly on this medium than on carbol-gelatine, but otherwise this appears to be the only advantage it possesses.

Stoddart's Method.—A novel method for the separation of the typhoid from the colon bacillus has recently been suggested by F. W. Stoddart (*Analyst*, May, 1897). This process is based upon the difference in the behaviour of the typhoid and coli bacilli when grown upon solid media incubated near their melting-points. The most satisfactory procedure is as follows: An agar-gelatine medium containing 0·5 per cent. of agar and 5 per cent. of gelatine, with the usual proportions of peptone and salt, is made, with all precautions to avoid loss of moisture during preparation. The reaction of the medium must be distinctly alkaline. The agar-gelatine is poured into flat-bottomed flasks or dishes to a depth of about 5 mm., sterilized, and allowed to cool slowly in the sterilizer, so as to avoid the exudation of moisture on the surface. The plates or flasks, which should not be more than a few days old, are then inoculated with a charged needle, and are incubated for twenty-four hours. In use, the centre of the medium is touched with a platinum loop charged with the material to be examined, the flask or dish is enclosed in a much larger one in order to prevent condensation, and the whole is placed in an incubator kept at 35° C. for twenty-four hours.

Pure cultures of typhoid so treated produce an opalescence occupying about two-thirds of the medium; *B. coli* gives a flat plate somewhat thicker and moister than the usual form. If the inoculation is made with both organisms, either from separate cultures or from a mixed culture, we get a flat plate of *B. coli* in the centre, with an opalescent halo of pure typhoid. Plates of this medium, inoculated direct from typhoid stools, gave without difficulty the same pure culture of typhoid bacilli; and it is anticipated that this will become a valuable diagnostic test, as easy of application, though not quite so rapid, as the serum test. It is best applied by putting two or three loops of stool into a little sterile broth, shaking and inoculating as described. Tap-water, also, inoculated with a trace of a broth culture of typhoid, or typhoid and coli mixed, readily yielded pure cultures of typhoid.

It was found, however, that, on applying it to polluted waters, in many cases there were present organisms, of

9

which more than a dozen were isolated, which responded
to all the accepted tests for the *B. coli*, but differed from it
in growing like typhoid on the agar-gelatine medium.
Moreover, when mixed cultures of typhoid and one or more
of these coli-like forms were inoculated on to this medium,
the typhoid was suppressed, and a pure culture of the non-
pathogenic organism obtained.

The same inhibition of typhoid also resulted when a
mixed culture in broth was attempted, though the true
B. coli and typhoid grow normally together. As these
organisms retain their vitality in media so highly carbolized
as to totally inhibit typhoid, there appears at present no
means of separating typhoid from them, unless the former
is present in such abundance as to be detectable by plate
culture, either in the usual form or as modified by
Elsner.

As soon as the colonies which develop on the carbolized
or potato-gelatine become sufficiently advanced, they are
examined with a lens, and any suspicious colonies are
carefully subcultured into faintly alkaline sterile milk-
tubes, which are then incubated at 37° C. for thirty-six
hours. The milk-tubes are then examined, and any that
have become coagulated are rejected as certainly not
typhoid.

Identification of the Typhoid Bacillus.

To establish the identity of an organism with the true
Eberth-Gaffky bacillus, a variety of tests must be applied;
and it is by no means certain, even at the present day, that
we are in possession of certain means of discrimination.

The tests that were regarded as satisfactory two or three
years ago are now largely discarded, and though many
fresh ones have been brought into use, the matter is very
far from being on a satisfactory basis.

No conclusions whatever can be drawn from the size of the organism when stained, as bacteria from cultures of different ages vary very considerably in size, and even in form, so that we have to rely entirely on appearances in cultures, on the chemical products of growth, etc.   In shape the typical typhoid bacillus from a twenty-four hours' culture is longer and more motile than the typical *B. coli*, and has long flagella in large numbers all along its body.   The following table gives the principal cultural means of distinction, which are generally considered of value :

Media.	Typhoid bacillus.	*Bacillus coli* group.
Gelatine plates.	No liquefaction. Colonies form large grayish-white expansions. Slow growth.	No liquefaction. The colonies are round and oval, with smooth-rimmed margins. Rapid growth.
Gelatine shake cultures.	No gas.	Copious gas formation.
Potatoes.	Slight, almost invisible growth.	Abundant slimy, yellowish growth.
Milk.	Turns faintly acid, but does not curdle.	Curdled in one to three days.
Broth.	No indol reaction.	Indol well marked in twenty-four to forty-eight hours.
25 per cent. nutrient gelatine at 37° C. (Klein).	Turbid.	Limpid, with strong pellicle after forty-eight hours.
Widal's serum reaction.*	Agglutination of bacilli.	No change.

Many other means of distinction have been proposed, such as growth on media containing an indicator; or, again, some observers recommend the test on animals. This latter test cannot be regarded as of much value, since some cultures of true typhoid are non-virulent to guinea-pigs, while some cultures of *Bacillus coli* are virulent.

* 'Applied Bacteriology,' Second Edition, pp. 178 *et seq.*

Isolation of B. Enteritidis Sporogenes (Klein).

Klein has shown reason to regard the *B. enteritidis sporogenes* discovered by him as diagnostic of sewage. It occurs in it in much smaller numbers than does *B. coli*, but it forms spores which are extremely persistent. It is pathogenic, producing very characteristic post-mortem appearances in animals, and causing acute diarrhœa in man. It is isolated by seeding a tube of 15 to 20 c.c. of sterile milk with the deposit produced by filtration of a large quantity of the water through a Pasteur filter-tube, heating for ten to fifteen minutes at 80° C., and incubating anaerobically (see p. 134) at 37° C. Within thirty-six hours characteristic changes occur in the milk, which are thus described by Klein (*L. G. B. Report*, 1897-98: Medical Supplement, p. 212):

'The characters of this change, which I have described in my former report (1895-96), are these: A large number of gas-bubbles in the upper layer of the milk (Fig. 6); bubbles, in addition, liberated in great abundance on shaking the tube, on exposing it to the light, or on sucking the fluid up into a pipette. The cream is torn or altogether disassociated by these gas-bubbles, so that the surface of the medium is covered with stringy pinkish-white masses of coagulated casein, enclosing a number of gas-bubbles; meanwhile, the main portion of the tube, formerly occupied by the milk, contains a colourless, thin, watery, clear or slightly turbid whey, with a few casein lumps adhering here and there to the sides of the tube; and at the bottom of the tube are small casein coagula. On opening the tube, the fluid within it has a distinct aromatic smell of butyric acid; and on testing the whey with blue litmus-paper, it is found that its reaction is distinctly acid. Under the microscope this fluid is found to contain numerous rods and cylinders, some connected in chains. Some rods are motile; others do not show motility. In aspect and in size the rods correspond to the *Bacillus enteritidis* described and figured in my former report (1895-96).

'The rapid production by *Bacillus enteritidis* of acid in milk culture is also shown by using, instead of ordinary milk, milk to which litmus has been added, such milk being, of course, blue, owing to its normal alkalinity. Incubating anaerobically an infected *litmus* milk culture, it is seen after a day or two that the milk has become completely deprived of its blue coloration. This is, of course, due to deoxidation or reduction of the litmus by the growth of the microbe. If the culture-tube is now taken out of the Buchner cylinder and exposed to the air, the culture reddens (by oxidation), the reddening beginning at the top of the culture fluid, and gradually extending into the depth, till all the fluid has become bright pink.'

These marked reactions are yielded also by the *B. butyricus* (Botkin), which, however, is not likely to occur in water, and is distinguished from the *B. enteritidis* by absence of pathogenic action. The bacillus of malignant œdema is distinguished from the *B. enteritidis* by, amongst other points, being much slower and less energetic in its action on milk, yielding a whey without acid reactions and butyric smell, and not staining on Gram's method. The bacillus of Rauschbrand is most readily distinguished by not staining by Gram's method, and not liquefying solidified blood-serum. The cultural differences appear equally after inoculation into animals, and are accompanied by marked difference of pathological effect.

The certainty with which this organism can be detected was shown by seeking it in sterile water to which very small quantities of sewage had been added; the organism was detected on inoculation into milk of one-half of the deposit on a Pasteur filter from 1,000 c.c. of sterile distilled water to which 0·002 c.c. of crude sewage had been added; that is to say, a pollution of 1 part of sewage in 500,000 of water was at once detected. When the composition of sewage is remembered, it will at once be evident that this result is far more delicate than could be obtained by chemical means; and, indeed, in a dilution of 1 part sewage in 20,000 of water the organism was found in less than one-third of the deposit from 1,000 c.c., while the saline and albuminoid ammonia and the oxygen absorbed were each nil. In other words, the pollution left the water, judged by any chemical standard, absolutely pure beyond what is known in nature, and at the same time the presence of this chemically-unde-tectable trace of sewage had introduced infective organisms. It is, therefore, to be hoped that in this organism, which is far more convenient for recognition than the *B. coli*, and which, through its strictly anaerobic character, is unlikely to occur outside sewage to anything like the extent of the *B. coli*, we may now have a far more direct symptom of sewage pollution than has yet been available.

In view of the especial importance which is to be attached to the pathological effects of this organism, it is practically indispensable that any cultures suspected of containing it should be sent to a medical bacteriologist for animal experi-ment.

Anaerobic Cultures.

It is necessary to point out, when investigating the characters of organisms in connection with sewage pollu-tion, that there exist in sewage and in water (particularly when the latter is polluted with manure or with sewage), in addition to the aerobic organisms, certain others which in the ordinary bacteriological analysis of water, sewage, and other materials are generally overlooked; these are organisms which grow only anaerobically, and which, therefore, do not make their appearance in the ordinary plate cultures. Although as a rule they are neglected in water and sewage analysis, they are, nevertheless, by their numbers and by their functions not without importance.

Many of the anaerobic bacilli are endowed with the power of more or less rapidly peptonizing and decomposing gelatine, of rapid multiplication in grape-sugar gelatine or grape-sugar agar in the depth, and of forming gas. Their spores can be heated to 80° C., for half an hour or an hour, without interfering with their subsequent power of germination into the bacilli.

In this connection, as has already been mentioned elsewhere, Klein traced an outbreak of severe diarrhœa to a sporogenous anaerobic bacillus, *B. enteritidis sporogenes*, which in morphological and cultural respects has some points in common with the *Bacillus butyricus*.

Furthermore, showing the importance of these organisms, it was recently stated by Klein (*Harben Lectures*, 1896) that he recently had occasion to examine a sample of sewage effluent which had been subjected to a certain treatment. This treatment affected the aerobic bacilli to such a degree that it reduced them from over 3,000,000 per cubic centimetre in the untreated effluent to 160 per cubic centimetre in the treated effluent; but when examining the effluent on the above lines for the presence of anaerobic spores, it was found that these latter, as also other spores, had remained unaffected by the treatment. This Klein considers is sufficient proof to show the importance of the analysis embracing also anaerobic cultures; the same applies to the bacteriological analysis of water, in which the detection of anaerobic sewage organisms may be difficult owing to their small number, but which may nevertheless contain the spores of some definite anaerobic sewage organisms.

A variety of different methods have been suggested from time to time for cultivating those bacteria which require anaerobic conditions. Perhaps the simplest plan is to use Esmarch 'roll cultures,' and either to place the tubes under a bell-jar and exhaust the air with a pump, or to connect the tubes with an apparatus for the production of hydrogen, and expel the air by a current of this gas. Buchner's method, which is also convenient, consists in placing the culture-tube within a larger test-tube, at the bottom of which is placed some pyrogallic acid; potassium hydrate solution is then added, and the large tube closed thoroughly with an indiarubber cork.

CONCLUSIONS.

The bacteriological results which are to be obtained in the manner above described must usually be taken, like all other factors of water examination, merely as so many hints. It must not be thought that these hints may be safely disregarded. On the contrary, they may furnish information which would on chemical examination be entirely wanting. We have added a drop of typhoid culture to sterile water, and diluted 1 c.c. of this diluted culture with 200 c.c. of ordinary tap-water which previously contained 200 organisms per c.c. The sample so obtained, estimated by gelatine plate, contained 900,000 organisms per c.c.; but the amount of pollution, measured chemically, was insufficient appreciably to raise the amount of albuminoid ammonia.* On the other hand, the absence of specific organisms furnishes no ground for passing as safe a water on which other means of examination, such as chemical analysis, have thrown suspicion. Take, for example, the case of a water suspected of having caused a typhoid outbreak. It is quite

* 'Applied Bacteriology,' Second Edition, p. 170. A still more direct illustration has been given above, p. 134.

possible that among the number of other organisms contained in the water a typhoid bacillus, even if present, might escape notice through the purely cultural difficulties which have been explained. But it is also possible that, while the water-supply in question did in fact cause the outbreak, the sample submitted for examination might be free from the organism. As has already been pointed out, water is not the natural habitat of pathogenic organisms; and while it is less inimical to their existence than air, and therefore for organisms which attack the digestive organisms a far more dangerous vehicle, the life of any pathogenic organism which falls into it is strictly limited if it does not happen to be attached to some substance peculiarly favourable to its existence. The incubation period of typhoid fever is about a fortnight; and in many cases therefore no doubt is cast upon the water, and no sample sent until three weeks after the specific infection. In this period the organism could well have disappeared. The cases where specific organisms are positively identified will be extremely rare; and, as has been pointed out, such cases should, where practicable, be remitted to the examination of a man who is experienced in their study. The feature above all others which is important, where it can be observed, is the variation from day to day. Certain variations are bound to occur, not only from climatic conditions, but also in many cases from local habits. The speed, for instance, at which water is withdrawn may mechanically alter the number of organisms which it carries with it. Systematic observation performed at regular short intervals under the same conditions will, however, rarely fail to reveal any variations in the number of bacteria, and in the same way will show in which of the three broad headings under which we look for them the increase or decrease occurs. Any substantial variations are to be taken as *primâ facie* ground for suspicion.

The investigations of Klein on the *B. enteritidis sporogenes* afford an excellent example of the absence of relation between the number of pathogenic organisms found in natural pollution and that of the ordinary aerobic organisms.

The bare numeration of bacteria in a single sample must plainly be stated to be of little value. Koch's 100 germs per c.c. is the conventional standard used principally for the purpose of being repudiated. Perhaps in unfiltered waters it may to this extent be read safely, that, when it is exceeded, extra vigilance for other signs of pollution may be desirable. In filtered waters this is still more the case; but strict conformity with it has been shown by Klein to be quite consistent with the presence of large animalcules which had passed through the filter-beds, and the consequent demonstration that at the time of sampling the filters had broken down, and the filtrate was polluted.

If organisms which, tested in the way described, appear to be *B. coli* or *B. enteritidis sporogenes* occur in large numbers, we have no hesitation in saying that on present knowledge they afford the strongest evidence of probable sewage pollution; but to an analyst whose experience renders him at all doubtful of his capacity rightly to identify either organism, we should recommend strongly the practice of first satisfying himself as to the identity of the suspected organisms with each other, and then seeking on a pure culture the confirmatory diagnosis of a bacteriologist; and in the case of the *B. enteritidis*, we should say that animal inoculation is absolutely indispensable. The same observations apply in general to organisms of specific disease, should he be so rarely fortunate as to discover one in water.

Bacteriological Examination of Filters.

When chemical analysis was the only means at command for examining water, it was found that in a majority of cases those waters which had been statistically convicted of spreading disease contained an excess of organic matter. Hence it was inferred that the organic matter was the cause of the disease; and filters were constructed of carbon, asbestos, natural stone, spongy iron, and similar materials for the purpose of removing this excess of organic matter. It was found that they all did so in a greater or less degree, but that their efficiency in this respect decreased on use, and ultimately disappeared until the filtering medium had been renewed or cleansed. With precisely similar results, preparations of these materials, such as silicated carbon, manganous carbon, magnetic iron, and the like, were tried for the same purpose, and many filters composed of successive strata of several of these were constructed. It ultimately became known that the diseases caused by water were due to micro-organisms, and that the presence of excess of organic matter in most waters which were dangerous was due to the fact that the microbe was generally either conveyed through excreta containing soluble organic substances, or best nourished in waters of such composition. The filters already in use were thereupon assumed to act by arresting the microbes contained in the water. This assumption was after a time supported by experiment, in which a small quantity of infected water was passed through the filter, and the filtrate was found to be sterile. Further investigation showed that this ceased to be the case when the filtration was continued for a few hours or less instead of a few minutes. It was found that in such case the filtrate contained the same organisms as the unfiltered water; and the sterility of the earlier filtrates was accordingly due to the circumstance that they had been examined before sufficient time had been allowed for the organisms to be washed through the filter. It was found that the chemical matter arrested by the filter only temporarily arrested a portion of the organisms, and often served as a suitable culture-ground for such organisms, which survived and multiplied for considerable periods in the filter before being ultimately washed through. In consequence, the number of organisms often became after a short time much larger in the filtrate than in the unfiltered water. Filters once polluted with the cholera or typhoid bacillus were also found to convey the bacillus to sterile water passed through them at considerable periods—up to six weeks or more—after pollution. This fact has been responsible for several epidemics, such as that of Lucknow in 1894, in which, out of 646 officers and men in the East Lancashire regiment, 143 were attacked by cholera, and 92 died. This epidemic was conclusively traced to the infection of the barrack-room filter by the cholera microbe.

Sand-filters.—A sand filter-bed consists of a layer of sand from 2 to 4 feet deep, supported on gravel. The fineness of the grains of sand, the depth of the filter, and the rate of filtration, all affect the working of the filter in the removal of organisms. The coefficients given for safe working are filtration through a sand-layer not less than 30 centimetres thick, at a rate not exceeding 100 millimetres per hour, and giving a filtrate containing not more than 100 bacteria per c.c. These coefficients, however, take no account of the class of sand used or character of water filtered, and they are no longer regarded as trustworthy. When a filter-bed is freshly constructed, organisms are washed through it with great rapidity, but after a certain quantity of water has passed through, or the water has been allowed to stand upon it for a certain time, a slimy coating of detritus and bacteria is formed on the surface. If water is slowly passed through the filter when this coating has been formed to a sufficient extent, which will occur after a period varying mainly with

the composition of the water, the majority of the bacteria
will be retained by this surface, either by sticking to it or
by being strained off.   The increasing thickness of this
coating will reduce the velocity with which the water passes,
and at the same time some of the bacteria will tend to grow
downwards into the lower strata of the filter, and, if the
process were continued long enough, would be washed
through into the filtrate, and ultimately become more
numerous there than in the unfiltered water.   The increasing resistance to the passage of water would also make it
necessary for the pressure to be increased, which would in
this class of filter assist the passage of organisms.   It is
therefore necessary in the working of sand-filters to run the
filtrate of each bed to waste, or to permit a body of water
to stand on the filter without filtration, until a sufficient
coating has been formed to arrest organisms; to stop
filtration when the deposit has increased to such extent as
to threaten the renewed passage of organisms; and to
remove the upper or filtering layer, and permit a fresh
deposit to be formed.   The indication for scraping usually
adopted is that the filter-bed no longer passes the required
quantity of water under the maximum permissible head.
The sufficiency of this practice has not been clearly shown.
No general rule can be given for the depth to which the
top layer must be removed, as it varies with the nature of
the water and sand, temperature, etc.

The ordinary rules for the selection of the epochs for
starting and arresting the filters, and the operation of
removing the upper surface, require considerable experience
and judgment, and it frequently happens that through
carelessness or unavoidable mistake the filtration is
imperfect.   Thus, in 1894 the filters at Nietleben, Altona,
Hamburg, and Stettin, being over 10 per cent. of the total
sand-filters in use in Germany, where great attention has

been given to the subject, passed the cholera organism, and permitted epidemics in their towns. In the same year the typhoid organism, of which the detection was difficult and uncertain with the means then at disposal, was, nevertheless, found beyond doubt in the Berlin water-mains.

Bacterial Filters.—It is obvious that, for the purpose of bacteriological investigation, such appliances as have been described are practically useless. Pasteur and Chamberland investigated a large number of earthen materials, beginning with ordinary biscuit porcelain. They found them to present very different degrees of resistance to the passage of bacteria. The difference did not appear to correspond to either the density of the material or the rate of filtration, in many cases a material of closer grain and less rapid output giving worse results than other materials more open in structure and more rapid in filtration. They ultimately found that the best results were obtained with a particular mixture of earths prepared with a special manipulation, and it is these substances which, when made in the well-known cylindrical form, constitute the Pasteur-Chamberland filter. This filter is found to be perfectly trustworthy in the removal of all organisms from liquids; it also retains any particulate matter, such as the fatty globules from milk.

The method of its action has not been determined, but it probably depends on some form of surface attraction, as many of the organisms which are arrested are considerably smaller than the pores of the material. It has been shown by repeated experiments that none of the many forms yet tried of biscuit earthenware, having practically the same appearance and analogous composition, possess the same efficiency as the Pasteur-Chamberland material; but no adequate reason has been discovered for the circumstance. A diagnostic test for the bacterial soundness of the Pasteur-

Chamberland tubes is to compress the air within them to an additional pressure of one-half to one atmosphere when the tube has been steeped in water, or is freshly taken from service. If held beneath a body of water, no air will escape from a sound tube, but a stream of bubbles will issue from any spot capable of passing bacteria. This test apparently does not apply to other forms of earthen filters, and for this reason they should not be used for the filtration of serum or during an epidemic, unless a portion of the filtrate is cultivated, and the bulk retained until it has been proved sterile. This applies particularly to filters made in the Pasteur-Chamberland form, in which a softer material, such as infusorial earth, is used, and a fresh filtering surface is accordingly exposed after each cleaning. Thus, the Berkefeld filter of infusorial earth, the tubes of which may initially be capable of preventing the direct passage of organisms, has a small portion of its outer surface removed each time it is cleaned. The consequence is that, sooner or later, a faulty surface is exposed, and the tube is liable to pass organisms even before the time when it is worn away sufficiently to break. The Pasteur-Chamberland tubes remain unaffected by cleaning or sterilization for an indefinite period. They may be sterilized by boiling water, or by saturated steam under pressure; or, alternatively, a suitable liquid disinfectant may be passed through them, with the advantage of dissolving at the same time the whole of the colloid substances deposited in its pores.

The Examination of Water-Filters.—The large majority of water-filters at present in use are incapable of preventing organisms from being washed through into the filtrate. In order to ascertain whether this is the case with any particular filter, it should be sterilized in the steam-sterilizer, and water containing known organisms should be passed through it for twenty-four hours. This water and the filter should during the time of the examination be maintained at a temperature below 5° C. This will almost invariably prevent any growth or multiplication of the organisms. Samples should be taken immediately after the filtration has begun, and at intervals during the day, and again at the end of twenty-four hours. If they are all sterile, the filter is capable of preventing organisms from being directly washed through. In the case of filters of very great density or depth of filtering medium it may be necessary to prolong the period of examination beyond the first day; but most ordinary filters which permit organisms to be washed through do so within the first few hours. It must be remembered that it is no advantage for a filter capable of permitting this passage of organisms to postpone it for a day or more, as the organisms will ultimately find their way into the filtrate, and in the meantime are likely in practical use to have increased in numbers. In the case of water-filters which withstand this test, and which may, therefore, be regarded as preventing organisms from being directly washed through, the further examination is a matter of some difficulty, and at the present time can only be conducted inferentially, or by comparison with a standard. The object of such examination is to discover whether pathogenic organisms in water can grow through the walls of the filter, and the difficulty in making the examination is that our information as to the circumstances which favour the multiplication of organisms in water, and which determine the maximum extent to which such multiplication may proceed in natural conditions, is quite incomplete. It is not possible to state of any given water whether it offers the maximum assistance to the growth of bacteria that may be found in natural water, or to say whether an organism under examination is capable of multiplying to the same extent as other specimens of the same organism might multiply in a natural water. In many researches, indeed, in which filters appeared to resist the penetration of organisms by growth it was not even certain whether the organisms under examination would grow in the water at all. The method which must, therefore, be employed is to take water containing known non-pathogenic organisms proved to multiply in it at suitable temperatures with sufficient freedom ultimately to penetrate the Pasteur-Chamberland tube, and to examine specimens of the filter, the efficiency of which is to be determined, at the same time, and with the same water-supply, as specimens of the Pasteur tubes themselves. The water must be kept at the optimum temperature, and the filtrates examined periodically. If the filter under examination retains the organisms for as long a time as the Pasteur, it must be considered as possessing the same efficiency. If, on the other hand, it passes the test-organisms before the Pasteur tube will do so, it is less efficient, and must for the present be considered insufficient for the prevention of infectious disease. There is an extremely large body of evidence to justify the conclusion that the resistance offered by the Pasteur-Chamberland tube is sufficient to prevent the passage of disease organisms from natural water. This evidence has been collected mainly in all parts of the French possessions, and published by the French Government; and since the filters have been introduced into this country and India, similar evidence has arisen. There is, however, no evidence to show that the resistance which it offers exceeds that which is necessary for affording trustworthy protection against water-borne disease. It is, therefore, not possible to accept any filter of less efficiency as affording a trustworthy guarantee against infection. In experiments of this kind care should be taken to procure several specimens of the filter under examination, and to ascertain that they fairly represent those intended for ordinary use. It is also desirable, when special test-organisms are artificially introduced, to avoid the simultaneous introduction of small quantities of culture material. It has been found that water and other fluids sterilized by heat may retain a toxic capacity, setting up, for instance, suppuration on inoculation into suitable animals; while the same liquid sterilized by filtration through a Pasteur-Chamberland tube produced no effect. At the present time these phenomena and the conditions which determine them are not sufficiently worked out to make it possible for filters to be adequately examined as to their capacity to produce similar results. A very full and interesting report, by Drs. Woodhead and Cartwright Wood, upon the efficiency of the various types of filters in use will be found in the *British Medical Journal*, vol. ii., 1894, pp. 1053, 1118, 1182, 1375, and 1486.

The examination of domestic water-filters by chemical analysis is not to be recommended. In the case of waterworks filters it has the great advantage, when pursued regularly at short intervals, of furnishing a means quite independent of the other methods of examination at disposal by which variations in quality can be detected; and so long as filters have to be used which permit a certain number of bacteria to pass into the filtrate, this class of examination is the best practicable. Domestic filters are not examined during use by their users, and for domestic purposes the use of filters possessing the efficiency of the Pasteur permits the use of a direct, instead of an indirect, criterion. The chemical properties of a filter, besides being in no way a guarantee of its protective capacity, are liable to change with use, and certain to disappear after a certain amount of use. There is, accordingly, a considerable risk of misleading a reader by describing the chemical properties of a filter; and, indeed, mistakes of this kind have notoriously been frequent.

Proper domestic filters are capable of giving the final and direct protection of preventing infective disease by arresting the organisms which cause it. While we may hope that ultimately the same may be said of our public filters, it must be taken as certain that for the domestic filter no less and no other criterion is now permissible. It is to be remarked that the chemical action of bacterial filters differs from that of the ordinary type, and as a rule is much less marked. Bearing in mind the fact that the chemical contents of water are in themselves innocuous, the presence in the filtrate, through a filter bacterially sound, of substances which would cast suspicion on a water not known to have been so purified, is absolutely without sanitary significance.

Filters of animal charcoal are sometimes used, with the object of removing lead. This action, so far as it occurs, depends on the presence of phosphates in the charcoal and the formation and arrest of insoluble lead phosphates. It follows, therefore, that when the phosphates are exhausted, the filter will be valueless for its purpose. In some cases this may not happen rapidly; but the presence of large quantities of calcium phosphate, which would be necessary to prevent it, encourages the growth of worms and other lower organisms, and filters containing this substance have, therefore, to be very frequently cleaned, sterilized, and recharged. For practical purposes no filter can, except under constant chemical supervision, be recommended for removing lead from water.

Sterilization of Water by Heat.

Various forms of apparatus have been devised to yield a continuous supply of water sterilized by heat. There are certain important points that must be kept in view in a successful apparatus for this purpose:

1. The temperature to which the water is raised, and the time that it is kept at that temperature, must suffice for absolute sterilization. This control should be automatically effected.

2. The apparatus must be so constructed that any deposit of lime-salts can be removed readily.

3. The apparatus should be so constructed that the gases naturally dissolved in water are not eliminated, and that the sterilized water is delivered within 2° or 3° of the original water, or the water will be flat and unpalatable.

The above requirements appear to be satisfactorily fulfilled by several appliances, such as the Equifex and the Maiche sterilizers. One of us had occasion recently to verify this in the case of the latter sterilizer; and there is no doubt that where water sterilized by heat is required, it can be obtained practically and without inconvenience. The methods of examination will be obvious from the statement of the conditions required.

It is perhaps too much to hope that our public water-supplies will, at any rate in the near future, be so treated as to deliver sterile water to the consumer; but in cases where water is employed in the preparation of aerated drinks, etc., it is not too much to expect of manufacturers that they should adopt special means to sterilize their water. This is more particularly to be insisted on in the case of products prepared from water (the ordinary town supply in many cases) which is not derived from a deep well, but from river or surface water.

VII. LAW RELATING TO WATER-SUPPLY.

THE supply of water in a district may be limited, and the quantity abstracted by the Water Authority of one district may affect the quantity available for the purposes of other districts deriving their supplies from the same source. The Public Health Act, 1875, therefore provides that no steps shall be taken by a Sanitary Authority which will injuriously affect the supply of water to which any person or body may be entitled; and in the frequent cases where a supply cannot be obtained without infringing on such rights, and where no existing water company provides the necessary supply, a Sanitary Authority desiring to provide water has to obtain a special Act. The provisions of such local Acts, and of the special Acts from which private water companies derive their powers, will not be discussed in this article.

The Public Health Act, 1875, makes various provisions for Authorities who desire to provide a supply from sources which can be tapped without interference with existing rights either of water companies or of water users. Such Authority cannot make the provision in any case where a company has Parliamentary powers for supplying the district unless the company is unable or unwilling to give a service proper and sufficient for the reasonable purpose specified by the Authority. Disputes as to the quality and sufficiency of the supply which the company can give, or as to the reasonableness of the Authority's demands, or as to the terms of supply so far as they are not determined by the company's Act, are to be settled by arbitration. Under the Waterworks Clauses Act, 1847, the company must bring their mains to any part of their district (*e.g.*, any street) on receiving a guarantee that the aggregate general rates shall be at least 10 per cent. of cost of providing and laying down the mains. In the absence of a company able and willing to give the reasonably required supply, the Sanitary Authority may, in an urban district, provide the district, or, say, part thereof, and in a rural district provide the district, or, say, contributory place therein, or, say, part of such contributory place with a supply proper and sufficient for public or private purposes. To this end they may construct and maintain waterworks, wells, etc., lease waterworks, or, with the sanction of the Local Government Board, purchase waterworks or water rights within or without the district, or contract with any person or body for a supply. They may acquire lands, leasements, and borrow money in the same way as for sewage works; and in the acquisition of the site of a well they acquire the right to draw all the water which percolates or is sucked into the well, provided it does not reach it through definite subterranean channels. Reservoirs exceeding 100,000 gallons in capacity may only be constructed by a Sanitary Authority if, after notice given, no objection is raised by an interested party, or if on such objection an inquiry is held by a Local Government Board inspector, and the construction, with or without modifications, allowed by the Board. Any two justices may make and enforce orders for securing the safety of any reservoir by a Sanitary Authority which they may be satisfied is in a dangerous condition. The powers of the Authority for carrying mains are the same as with sewers. They must give three months' notice by local advertisement before beginning the work, detailing the nature and course proposed for the mains, and naming a place where a plan can be seen, and copies of such notice must be served on all owners or reputed owners, lessees or reputed lessees, or occupiers of lands through, across, under, or on which the mains are to be carried, and on the overseers of the parishes in which they lie, and on the trustees, highway surveyors, and others having the care of the roads or streets affected. If objection be taken by any of these parties, or by any other interested owner, lessee, or occupier, and notice of such objection is served upon the Sanitary Authority within the three months, and is not withdrawn, the works require the preliminary sanction of the Local Government Board, after local inquiry and report by their inspector. Alterations in existing mains may be made, provided the works do not constitute a nuisance and no person is left without a supply who is entitled to it. A map of the mains may be made, and, if made, must be kept at the office, open to inspection by ratepayers at all reasonable hours. Erection of buildings over mains without consent of the Sanitary Authority subjects the offender to a fine of £5, and 40s. a day during continuance of the offence, power being also reserved to the Authority to remove such buildings and summarily recover the expense from the offender. An Urban Sanitary Authority or a Rural Sanitary Authority having the powers of a Highway Board, when the mains are laid outside its district, may not break up streets until notice has been given to the persons controlling them, and, if required by such persons, in accordance with a plan agreed by them, or, in default of agreement, determined by two justices, and under the superintendence of the persons, or their officer, if he attend for the purpose, and subject to reasonable expedition in reinstating the streets.

Existing public cisterns, pumps, wells, reservoirs, conduits, and other works, for gratuitous supply of water, belong to, and are under the control of, the Sanitary

Authority, who may maintain and keep them supplied with pure and wholesome water, and may, with like powers, substitute for them other works equally convenient; and, subject to the provisions of the Act, may construct other works for gratuitous supply of water to inhabitants who choose to fetch it—not for sale, but for private use.

In any works made or bought by the Sanitary Authority they are bound to keep a pure and wholesome supply of water, and they are at liberty, but not bound, to provide constant service, under pressure sufficient to carry the water to the top of the highest house in the district. A Sanitary Authority may, with sanction of the Local Government Board, on terms to be agreed or decided by arbitration, as provided in the Public Health Act, 1875, provide a supply for a neighbouring Authority. They may charge for supply by a rate assessed on the net annual value of the premises, as in the case of a general district rate. This rate is payable quarterly, in advance, and in the case of a house of less annual value than £10 is payable to the Authority by the owner. In the alternative, they may agree on a water rent with individuals supplied, and, in particular, may agree to supply by meter, the rents or charges being recoverable in the same way as rates. Under the Public Health (Water) Act, 1878, any Sanitary Authority may be compelled to exercise these powers on the demand of ten ratepayers in an urban district, or in the case of a rural district by five ratepayers, in a contributory place; and such rates or rents may be charged on all dwelling-houses within 200 feet of any standpipe provided by the Authority, unless such house has another sufficient source of wholesome water within reasonable distance, and does not use the water of the Sanitary Authority.

A Sanitary Authority may provide water for public baths or wash-houses, or for trade purposes, on terms to be agreed between them and the consumers, and may construct works for gratuitous supply of public baths or wash-houses, whether worked for private profit or supported out of rates.

Powers for proceeding against persons polluting water-courses within or without the district of the Authority, and for recovering penalties, are provided by the Public Health Act, 1875. Further powers are given in regard to streams within or passing through or by the district by the Rivers Pollution Prevention Act, 1876. A summary penalty of 40s. may be recovered against persons throwing or allowing to be thrown or placed in a watercourse within the district of a Sanitary Authority which has adopted Part III. of the Public Health Amendment Act, 1890, any filth, rubbish, or other substance likely to cause annoyance.

The waste or misuse of water is prohibited under penalty by the Public Health Act, 1875. Under the same Act the inhabitants may, after fourteen days' notice to the Sanitary Authority, and with the consent of owners or occupiers of any intervening lands, lay surface-pipes from their premises to the mains, and, after two days' notice, make the necessary connection to the mains, under the super-intendence and direction of the surveyor or other officer of the Authority, if he see fit to attend. For this purpose the pavement may be broken up, but must be made good.

If it is represented to the Sanitary Authority by any person that in their district any public or private well, tank, cistern, or pump used or likely to be used for human drinking or domestic purposes, or for making human drink, is so polluted as to be injurious to health, the Authority may, under the Public Health Act, 1875, summarily proceed against the owner, if private, or against any person alleged to be interested in the supply; if public, for an order for permanent or temporary closing, restriction of use to certain purposes, or otherwise, as may be thought necessary to prevent injury to health. The court may order an analysis at the cost of the Sanitary Authority, but the analysis need not be made at Somerset House as provided in the Food and Drugs Acts. On failure to comply with such an order, the court may authorize the Sanitary Authority to do the necessary work, and summarily recover the cost from the person on whom the order is made.

The Sanitary Authority must, on their Surveyor's report that a house within their district has no proper water-supply, and that one could be provided within the limit of cost provided by the Act, serve notice on the owner to provide such supply within a specified time, and, in default, may themselves, or by the water company, lay on such supply and recover the ordinary rates, and may either summarily recover from the owner the expense incurred in such works, or may declare them to be private improvement expenses—*i.e.*, expenses of which the payment may be spread over a short term of years with interest, and recovered against the occupier as well as the owner. The limit of cost is the rates provided under the local Act, if there is one ; and if there is not, it is either twopence per week, or such other sum as the Local Government Board may in each case decide, or within such general tariff as, under the Public Health (Water) Act, 1878, the Local Government Board may, on any such application, order to apply generally throughout the district, or any part of it. Where capital outlay is required to an extent making it impossible to come within the limits of cost, a Rural Authority may enforce provisions of the Public Health (Water) Act, 1878, which put the duty of providing a supply on the owner. If the same case arises in an urban district, the Local Government Board may invest the Urban Sanitary Authority with like powers unconditionally, or to such extent and on such terms as they think fit. If it is impossible for either owner or Authority to provide the supply, and the absence makes the house a nuisance or injurious to health, proceedings may be taken to obtain a justice's order to close it. When an Urban Sanitary Authority fail to provide a proper water-supply, and it could be provided at a reasonable cost, the Local Government Board must, under the Public Health Act, 1875, on the complaint of any individual, inquire into the matter, and, if satisfied that default has been made, order the Authority to perform the work within a specified time. In default, the order may be enforced by mandamus, or the Local Government Board may appoint a person to perform the service, and direct the expenses to be paid by the Authority in default. In the case of a Rural District Council, the County Council may, under the Local Government Act, 1894, on the complaint of the Parish Council, or, in its absence, of the parish meeting, inquire, and, if satisfied of the default, may transfer the powers of the District Council to themselves and execute the works, or may exercise the same powers as the Local Government Board exercise in regard to Urban Sanitary Authorities. These provisions are those provided under the respective Acts for default in provisions of sewers. Without prejudice to the obligations on the Rural Sanitary Authority to provide a proper supply, the Parish Council may, under the Local Government Act, 1894, utilize any well, spring, or stream within their parish, and, subject to the rights of third parties, provide facilities for obtaining water therefrom.

Rural Sanitary Authorities may, under the Public Health (Water) Act, 1878, cause periodical inspections to be made, and incur reasonable expenses therefor. They, or their officer, or other person appointed for the purpose, have the same right of entry into houses reasonably suspected to be without proper water-supply as their officer has in the case of a suspected nuisance, and the same penalties for refusing admission apply. Where, on the report of their Inspector of Nuisances or Medical Officer of Health, it appears to the Authority that a house in their district is unprovided with a proper water-supply, and they are of opinion that one can be provided at a cost not exceeding a capital sum on which interest at £5 per cent. per annum would amount to twopence a week, or to threepence a week if fixed by the Local Government Board on the application of the Authority, and that the cost ought to be charged as private improvement expenses, the Authority may, by notice served on the owner, order a supply to be provided within a specified time, not exceeding six months. In case of default, and of their seeing no reason to withdraw or modify the notice, they may serve a second notice, giving the owner a further month, and warning him that if the notice is not complied with the Authority will themselves execute the work, the cost of which will be either payable by the owner, or as a private improvement expense; and in case of default, and of their seeing no reason to withdraw or modify the notice, the Authority may themselves execute the work under the same powers until the works are completed, as they would have in the case of a nuisance, and may, after completion of the work, either recover the expense summarily, or declare it to be a private improvement expense. Where two or more houses are in default, the Authority may, if entitled to do the works of the houses separately, and no greater expense would be incurred, provide a joint supply and apportion the expense. In the case of such proceedings, the owner may, on any or all of the grounds that either the supply is not required, or the specified time for providing it is too short, or it cannot be provided at reasonable cost, or the Authority ought themselves to provide or purify a supply for the district or contributory place, or the cost, or part of it, ought to be a charge on the district or contributory place, within twenty-one days from service of the second notice address a memorial to the Authority stating his objections. If the objections do not allege liability of the Authority to provide or purify the supply, or of the district or contributory place to defray all or part of the cost, the Authority must, before proceeding with the work, apply to a court of summary jurisdiction, which, after hearing the owner, may authorize the Authority to proceed with the work. If either or both of those grounds are alleged, the work is stayed for decision of the Local Government Board on the memorial ; and the Board may confirm the order, with or without modifications, or may apportion the expense between the owner and the Authority of the contributory district, or between the owner and any other person. The owner of a house between whom and another owner, or other owners, the Authority has apportioned the expense of a joint supply, must receive notice from the Authority, and may within twenty-one days from service apply to a justice for revision of the apportionment.

Under the Public Health Act, 1875, no house erected or completely rebuilt from the ground-floor since July 4, 1878, can be occupied without a certificate from the Sanitary Authority that there is a proper water-supply within a reasonable distance. The owner may appeal to a court of summary jurisdiction. The Authority may summarily recover a penalty, not exceeding £10, against the owner who occupies a house or allows it to be occupied despite of their order, unless on appeal the court has authorized the house to be inhabited.

The water-supply of London is the subject of the special

Acts of 1857 and 1871, as well as of the local Acts of the eight companies concerned, and further provisions are made by the Public Health (London) Act of 1891. Some of the special provisions of the Metropolis Water Acts are so exceptional as to deserve mention. The Local Government Board audits, and to some extent corrects, the accounts of the companies, and periodically examines their water; their approval is necessary before any new source of supply can be utilized; and if the use has been authorized by Parliament, it is subject to the Local Government Board's certificate that the conditions of the Authorizing Act have been complied with. They may inquire into complaints as to quality and quantity of supply, and require their remedy by the company; and their approval is necessary to the regulations of the companies as to abuse, waste, or contamination of the supplies. On the failure of a company to repeal or alter any such regulation after request, made in writing by either the County Council or any ten consumers, the Local Government Board may, after report of an expert, repeal or modify such regulation. Notice of all regulations must be given to the County Council within four days of its being made, and the Board must, if desired, hear the County Council before confirming them. The County Council may demand a constant supply in any district, or may object to a company's proposal to give one, and on a memorial by either party within one month setting forth objections, the Local Government Board may make inquiry, and, after hearing both parties, make such order on the subject as they think right, subject only to its being established that the regulations touching abuse and pollution of supply have been made and put into operation in the district, and to the company not establishing that at any time after two months from service of the demand for constant supply more than one-fifth of the premises in the

district were unfitted with the prescribed fittings. In the
latter case the County Council may require the owner or
occupier to provide the prescribed fittings within a specified
time, and in default may provide or repair them, and
recover the cost from the person liable to pay the water
rate, or on whose credit the water is supplied, or from the
owner. The Local Government Board may also of its own
motive require a constant supply to be provided in any
district if it is satisfied that the County Council ought, but
does not, or unreasonably delays to do so, or that, through
insufficiency of supply or imperfect storage, the health of
the district is, or is likely to be, injured.

The Public Health (London) Act, 1891, imposes duties on
the Sanitary Authorities, including Vestries and District
Boards.   An occupied house without proper water-supply,
or without the fittings of which the absence is a nuisance
within the Metropolis Water Act, 1871, is a nuisance liable
to be dealt with summarily, and, if a dwelling-house, is unfit
for human habitation.   Houses built or rebuilt after the
date of the Act come under similar provisions to those
applying to the rest of the county under the Public Health
Act, 1875, the appeal lying to Petty Sessions, and a fine of
20s. per day for continuing default being added to the £10
penalty.   Notice must be given within twenty-four hours to
the Sanitary Authority of any case in which they have law-
fully cut off the supply from an inhabited dwelling-house.
The Sanitary Authority must make by-laws, which must be
submitted within six months for the sanction of the Local
Government Board, for safeguarding tanks and other
vessels for storing water used, or likely to be used, by men
for drinking or domestic purposes, or for the manufacture
of human drink.   The Model By-laws of the Local Government Board provide that such vessel must be emptied and
cleaned at least once in six months, and oftener if necessary

for cleanliness and freedom from pollution; if outside the building, or otherwise exposed, is to be made and kept covered; and, in case of common use of any such vessel by two or more tenants of the same premises, the obligation to comply with these by-laws is on the owner. Public cisterns, reservoirs, wells, pumps, fountains, and works for gratuitous supply of water which are not vested in any other Authority, are vested in the Sanitary Authority, and the provisions of the Act are similar to those of the Public Health Act, 1875. The Authority has also power to provide and maintain fresh public wells, pumps, and fountains, and, in addition to any penalties which may be applicable on demand, payment of the cost of making good any damage wilfully done to such works or any part of them. The provisions for dealing with polluted wells, tanks, etc., include similar powers to those of the Public Health Act, 1875, the judicial authority being the Petty Sessions. In addition, both assign probability of pollution and danger to health as a ground for closing them, where the former Act demanded that the supply should be actually polluted and injurious to health, and gives no power to the court to order restricted use. The penalty for disobedience is a sum not exceeding £20.

For fouling or polluting any well, pump, etc., a penalty of £5, with a further 20s. per day during continuance of offence, is prescribed.

VIII. AERATED TABLE WATERS.*

THE improvements in the manufacture of plant for making aerated waters has led to a great increase in the production, both on a large scale and at local factories. It is fairly certain that a considerable part of this increase is due to the impression that such waters are safer than ordinary water; and, indeed, the reputation in this respect of the aerated waters from some districts has led to a large demand for them in even distant parts of the kingdom. The general feeling that aerated waters are safer than non-aerated is correct to this extent, that the much smaller supply required by an aerated water factory can be provided under precautions which are never taken with the much larger volume of a public supply intended for all purposes. In recognising this advantage, however, it must be clearly remembered that the manipulation involved in bottling aerated waters introduces an element of risk which does not occur in the public supply. A case in point, which occurred recently, was that of a manufacturer who, becoming unwilling to use a polluted well for manufacturing aerated waters, proposed to do so for washing out the bottles, a course which would have been entirely incompatible with the purity of the water, and was, in fact, abandoned on the advice of the analyst. Reputations for

* By the courtesy of the proprietors of the *Medical Press*, we are permitted to reprint this article, which appeared in that journal in the issue of January 11, 1899.

11

purity, on the other hand, have over and over again been shown to be entirely untrustworthy. As an illustration of this it may be mentioned that in the town of Maidstone there was a thriving industry of old standing in mineral waters, based on a reputation for purity; and this firm continued to export their products to all parts of the kingdom until the actual outbreak of the great epidemic.

The liability of aerated waters to convey infective disease has been thoroughly recognised for many years. The presence of carbonic acid in water, if it exercises any disinfectant effect, does so only after the lapse of considerable periods of time, and cannot be taken into serious account as a means of affording protection to the public. The results which we are about to publish are intended to indicate rather the natural character of the water which is used in the brands under examination than the precise merit of the finished product. It cannot be too strongly recognised that the first line of defence against infection is the purity of the water-supply. The purity is in almost all cases liable by accident to be impaired, and the demand for the adoption of uniform and trustworthy artificial means of bacterial purification, such as Pasteur filtration, has been increasing, and, in our opinion, ought to increase until such means are universally adopted. In this way the public will obtain, and indeed has already begun to obtain, a second line of defence, and when it is uniformly adopted the impression that aerated waters are usually safer than the waters of a town supply will, for the time being, be justified.

'Mineral waters' are occasionally examined by public analysts under the Sale of Food and Drugs Act.

We are only able to find records for one year—namely, 1893, in which 168 samples were examined, and 32, or 19 per cent., were adversely reported on.

Unfortunately, there does not seem to be any published record giving the reasons in all cases for their condemnation, the report only stating that in some cases lead was detected in small quantities.

'CAMWAL' SODA-WATER.

The Chemists Aerated Mineral Waters Association, Limited, have factories in different parts of the country (London, Manchester, Birmingham, Bristol, Harrogate, and Mitcham), at which they prepare mineral waters. The samples we have examined were prepared from the 'New River' water. Our analyses also included samples of the latter.

Label—Camwal Soda-Water.

PHYSICAL CHARACTERS.

| Colour ... ... Faint blue. | Taste ... ... Normal. |
| Smell (by Boudriment's method) ... Normal. | Suspended matter ... None. |

CHEMICAL CHARACTERS.

	Grains per Gallon.		Grains per Gallon.
Total solids (dried at 120° C.)	37·0	Saline ammonia ...	Traces.
		Albuminoid ammonia	None.
Mineral solids (recarbonated)	—	Oxygen absorbed (in 15 minutes) ...	—
Loss on ignition ...	—	Oxygen absorbed (in 4 hours)	Traces.
Chlorides (as chlorine)	1·5		
Hardness (total) ...	16·0	Poisonous metals ...	None.
Nitrites	None.	Phosphates	None.
Nitrates (as nitrogen)	·27		

Label—New River Water (used by 'Camwal').

PHYSICAL CHARACTERS.

| Colour ... ... Faint blue. | Taste ... ... Normal. |
| Smell (by Boudriment's method) ... Normal. | Suspended matter ... None. |

11—2

CHEMICAL CHARACTERS.

	Grains per Gallon.		Grains per Gallon.
Total solids (dried at 120° C.)	24·6	Saline ammonia ...	·008
Mineral solids (recarbonated)	18·2	Albuminoid ammonia	Traces.
		Oxygen absorbed (in 15 minutes) ...	—
Loss on ignition ...	6·4	Oxygen absorbed (in 4 hours)	None.
Chlorides (as chlorine)	1·5		
Hardness (total) ...	12·5	Poisonous metals ...	None.
Nitrites	None.	Phosphates	None.
Nitrates (as nitrogen)	·275		

THE IDRIS MINERAL WATER COMPANY,

Pratt Street, Camden Town.

The water used is derived from a deep well, and for the sake of comparison we examined the soda-water and the well water also, the figures obtained being given below:

Label—Idris Soda-Water.

PHYSICAL CHARACTERS.

Colour	Faint blue.	Taste	Normal.
Smell (by Boudriment's method) ...	Normal.	Suspended matter ...	None.

CHEMICAL CHARACTERS.

	Grains per Gallon.		Grains per Gallon.
Total solids (dried at 120° C.)	73·5	Saline ammonia ...	0·0028
Mineral solids (recarbonated)	—	Albuminoid ammonia	Traces.
		Oxygen absorbed (in 15 minutes) ...	—
Loss on ignition ...	—	Oxygen absorbed (in 4 hours)	Traces.
Chlorides (as chlorine)	9·8		
Hardness (total) ...	8·0	Poisonous metals ...	None.
Nitrites	Absent.	Phosphates	None.
Nitrates (as nitrogen)	Traces.		

Label—Water from Idris Well.

PHYSICAL CHARACTERS.

Colour	Faint blue.	Taste	Normal.
Smell (by Boudriment's method) ...	Normal.	Suspended matter ...	None.

CHEMICAL CHARACTERS.

	Grains per Gallon.			Grains per Gallon.
Total solids (dried at 120° C.)	41·0	Saline ammonia ...	None.	
Mineral solids (recarbonated)	35·0	Albuminoid ammonia	None.	
Loss on ignition ...	6·0	Oxygen absorbed (in 15 minutes) ...	—	
Chlorides (as chlorine)	8·0	Oxygen absorbed (in 4 hours)	None.	
Hardness (total) ...	4·0	Poisonous metals ...	None.	
Nitrites	None.	Phosphates	None.	
Nitrates (as nitrogen)	Traces.			

MESSRS. ELLIS AND SON,

Mineral Water Manufacturers, Ruthin, Wales.

The water used is derived from a deep spring. The figures we have obtained on the water are given below:

Label—Ellis's Soda-Water.

PHYSICAL CHARACTERS.

Colour	Faint blue.	Taste	Normal.
Smell (by Boudriment's method) ...	Normal.	Suspended matter ...	None.

CHEMICAL CHARACTERS.

	Grains per Gallon.			Grains per Gallon.
Total solids (dried at 120° C.)	86·6	Saline ammonia ...	Traces.	
Mineral solids (recarbonated)	—	Albuminoid ammonia	·0056	
Loss on ignition ...	—	Oxygen absorbed (in 15 minutes) ...	—	
Chlorides (as chlorine)	1·0	Oxygen absorbed (in 4 hours)	·012	
Hardness (total) ...	10·0	Poisonous metals ...	None.	
Nitrites	None.	Phosphates	None.	
Nitrates (as nitrogen)	0·14			

Label—Messrs. Ellis's Spring, Ruthin.

PHYSICAL CHARACTERS.

Colour	Faint blue.	Taste	Normal.
Smell (by Boudriment's method) ...	None.	Suspended matter ...	None.

CHEMICAL CHARACTERS.

	Grains per Gallon.		Grains per Gallon.
Total solids (dried at 120° C.)	14·0	Saline ammonia ...	Traces.
		Albuminoid ammonia	Traces.
Mineral solids (recarbonated)	11·0	Oxygen absorbed (in 15 minutes) ...	Traces.
Loss on ignition ...	3·0	Oxygen absorbed (in 4 hours)	·01
Chlorides (as chlorine)	1·6		
Hardness (total) ...	9·0	Poisonous metals ...	None.
Nitrites	None.	Phosphates	None.
Nitrates (as nitrogen)	·15		

MESSRS. ROSS AND SONS, LIMITED,

Mineral Water Manufacturers, Belfast and London.

The following figures were obtained on a sample of Messrs. Ross and Sons' 'double soda-water':

Label—Ross's Double Soda-Water.

PHYSICAL CHARACTERS.

Colour	Faint blue.	Taste	Normal.
Smell (by Boudriment's method) ...	Normal.	Suspended matter ...	None.

CHEMICAL CHARACTERS.

	Grains per Gallon.		Grains per Gallon.
Total solids (dried at 120° C.)	187·0	Saline ammonia ...	·0056
		Albuminoid ammonia	Traces.
Mineral solids (recarbonated)	—	Oxygen absorbed (in 15 minutes) ...	—
Loss on ignition ...	—	Oxygen absorbed (in 4 hours)	·043
Chlorides (as chlorine)	26·0		
Hardness (total) ...	0·5	Poisonous metals ...	None.
Nitrites	None.	Phosphates	None.
Nitrates (as nitrogen)	Traces.		

Label—Well Water used by Ross and Sons.

PHYSICAL CHARACTERS.

Colour	Faint blue.	Taste	Normal.
Smell (by Boudriment's method) ...	None.	Suspended matter ...	None.

CHEMICAL CHARACTERS.

	Grains per Gallon.			Grains per Gallon.
Total solids (dried at 120° C.)	17·0	Saline ammonia ...		·002
Mineral solids (recarbonated)	12·0	Albuminoid ammonia		Traces.
		Oxygen absorbed (in 15 minutes) ...		None.
Loss on ignition ...	5·0	Oxygen absorbed (in 4 hours)		Traces.
Chlorides (as chlorine)	6·1			
Hardness (total) ...	—	Poisonous metals ...		None.
Nitrites	None.	Phosphates		None.
Nitrates (as nitrogen)	Traces.			

MESSRS. SCHWEPPE AND CO.,

Berners Street, London.

The following figures were obtained on a sample of soda-water prepared by Messrs. Schweppe and Co. :

Label—Schweppe's Soda-Water.

PHYSICAL CHARACTERS.

Colour	Faint blue.	Taste	Normal.
Smell (by Boudriment's method) ...	Normal.	Suspended matter ...	None.

CHEMICAL CHARACTERS.

	Grains per Gallon.			Grains per Gallon.
Total solids (dried at 120° C.)	66·2	Saline ammonia ...		0·028
Mineral solids (recarbonated)	—	Albuminoid ammonia		0·007
		Oxygen absorbed (in 15 minutes) ...		—
Loss on ignition ...	—	Oxygen absorbed (in 4 hours)		0·043
Chlorides (as chlorine)	4·3			
Hardness (total) ...	19·0	Poisonous metals ...		None.
Nitrites	None.	Phosphates		None.
Nitrates (as nitrogen)	0·05			

Label—Well Water used by Schweppe's, Limited.

PHYSICAL CHARACTERS.

Colour	Faint blue.	Taste	Normal.
Smell (by Boudriment's method) ...	None.	Suspended matter ...	None.

CHEMICAL CHARACTERS.

	Grains per Gallon.			Grains per Gallon.
Total solids (dried at 120° C.)	25·0	Saline ammonia ...		·001
Mineral solids (recarbonated)	20·0	Albuminoid ammonia		None.
		Oxygen absorbed (in 15 minutes) ...		—
Loss on ignition ...	5·0	Oxygen absorbed (in 4 hours)		·05
Chlorides (as chlorine)	4·2			
Hardness (total) ...	12·5	Poisonous metals ...		None.
Nitrites	None.	Phosphates		None.
Nitrates (as nitrogen)	Traces.			

'SALUTARIS.'

As will be seen from the figures below, 'Salutaris' is prepared from a distilled water. Sample 'A' was furnished us by the manufacturers, while sample 'B' was purchased by us without their knowledge:

Physical Characters.	'A.'	'B.'
Colour	Faint blue.	Faint blue.
Taste	Normal.	Normal.
Suspended matter	Absent.	Absent.

Chemical Characters.	Grains per Gallon.	
Total solids	2·6	2·0
Loss on ignition	1·2	0·6
Mineral solids	1·4	1·4
Chlorine	0·2	0·2
Hardness	1·0	1·0
Nitrites	Traces.	Traces.
Nitrates and poisonous metals	None.	None.
Saline ammonia	·0028	·0056
Albuminoid ammonia ...	·0084	·0084
Oxygen absorbed	·055	·056

From the results given above we conclude that in all cases the water employed in manufacture and the finished products are of satisfactory organic purity, and free from any trace of metallic contamination.

INDEX.

www.ingramcontent.com/pod-product-compliance
Lightning Source LLC
Chambersburg PA
CBHW021804190326
41518CB00007B/446